AF806333

# Blockchain and IoT

This book offers a detailed exploration of the transformative potential of the Internet of Things (IoT) and blockchain in reshaping industries. Through a combination of theoretical foundations, practical case studies, and innovative applications, it highlights how these cutting-edge technologies can revolutionize business models, enhance operational efficiency, and foster innovation. This book is a helpful guide for scholars, researchers, and professionals engaged in research on the Internet of Things, artificial intelligence, machine learning, blockchain, and their applications at the industry level.

- Explains how IoT devices can leverage blockchain technology to enhance security, transparency, and efficiency in data management and transactions.
- Explores the security challenges faced by IoT devices and how blockchain can address these challenges by providing a decentralized and tamper-resistant platform for data exchange and storage.
- Investigates the potential of cloud-based IoT platforms, analytics, protocols, and business models to revamp the workforce.
- Discusses how companies can leverage the data generated by IoT devices securely through blockchain-based data marketplaces.
- Provides real-world examples and case studies of successful implementations of IoT and blockchain technologies across different industries such as healthcare, manufacturing, logistics, and energy.

# Chapman & Hall/CRC Internet of Things: Data-Centric Intelligent Computing, Informatics, and Communication

The role of adaptation, machine learning, computational Intelligence, and data analytics in the field of IoT Systems is becoming increasingly essential and intertwined. The capability of an intelligent system is growing depending upon various self-decision-making algorithms in IoT Devices. IoT based smart systems generate a large amount of data that cannot be processed by traditional data processing algorithms and applications. Hence, this book series involves different computational methods incorporated within the system with the help of Analytics Reasoning, learning methods, Artificial intelligence, and Sense-making in Big Data, which is most concerned in IoT-enabled environment.

This series focuses to attract researchers and practitioners who are working in Information Technology and Computer Science in the field of intelligent computing paradigm, Big Data, machine learning, Sensor data, Internet of Things, and data sciences. The main aim of the series is to make available a range of books on all aspects of learning, analytics and advanced intelligent systems and related technologies. This series will cover the theory, research, development, and applications of learning, computational analytics, data processing, machine learning algorithms, as embedded in the fields of engineering, computer science, and Information Technology.

***Series Editors:***

**Data Science for Effective Healthcare Systems**
*Hari Singh, Ravindara Bhatt, Prateek Thakral and Dinesh Chander Verma*

**Internet of Things and Data Mining for Modern Engineering and Healthcare Applications**
*Ankan Bhattacharya, Bappadittya Roy, Samarendra Nath Sur, Saurav Mallik and Subhasis Dasgupta*

**Energy Harvesting: Enabling IoT Transformations**
*Deepti Agarwal, Kimmi Verma and Shabana Urooj*

**SDN-Supported Edge-Cloud Interplay for Next Generation Internet of Things**
*Kshira Sagar Sahoo, Arun Solanki, Sambit Kumar Mishra, Bibhudatta Sahoo and Anand Nayyar*

**Internet of Things: Applications for Sustainable Development**
*Niranjan Lal, Shamimul Qamar, Sanyam Agarwal, Ambuj Kumar Agarwal and Sourabh Singh Verma*

**Artificial Intelligence for Cognitive Modeling: Theory and Practice**
*Pijush Dutta, Souvik Pal, Asok Kumar and Korhan Cengiz*

**Cognitive Predictive Maintenance Tools in Brain Diseases: Design and Analysis**
*Shweta Gupta*

**Cognitive Computing for Smart Automotive Transportation: Technology and Applications**
*By G. R. Kanagachidambaresan, Archana Naganathan, Niresh Jayarajan and Sheldon S. Williamson*

**Blockchain and IoT: Foundations, Applications and Case Studies**
*By R Anandan, D Akila and Souvik Pal*

# Blockchain and IoT

## Foundations, Applications and Case Studies

Edited by
R Anandan, D Akila, and Souvik Pal

CRC Press
Taylor & Francis Group
Boca Raton London New York

CRC Press is an imprint of the
Taylor & Francis Group, an **informa** business
A CHAPMAN & HALL BOOK

Designed cover image: ShutterStock

First edition published 2025
by CRC Press
2385 NW Executive Center Drive, Suite 320, Boca Raton FL 33431

and by CRC Press
4 Park Square, Milton Park, Abingdon, Oxon, OX14 4RN

*CRC Press is an imprint of Taylor & Francis Group, LLC*

© 2025 selection and editorial matter, R Anandan, D Akila, and Souvik Pal; individual chapters, the contributors

Reasonable efforts have been made to publish reliable data and information, but the author and publisher cannot assume responsibility for the validity of all materials or the consequences of their use. The authors and publishers have attempted to trace the copyright holders of all material reproduced in this publication and apologize to copyright holders if permission to publish in this form has not been obtained. If any copyright material has not been acknowledged please write and let us know so we may rectify in any future reprint.

Except as permitted under U.S. Copyright Law, no part of this book may be reprinted, reproduced, transmitted, or utilized in any form by any electronic, mechanical, or other means, now known or hereafter invented, including photocopying, microfilming, and recording, or in any information storage or retrieval system, without written permission from the publishers.

For permission to photocopy or use material electronically from this work, access www.copyright.com or contact the Copyright Clearance Center, Inc. (CCC), 222 Rosewood Drive, Danvers, MA 01923, 978-750-8400. For works that are not available on CCC please contact mpkbookspermissions@tandf.co.uk

*Trademark notice*: Product or corporate names may be trademarks or registered trademarks and are used only for identification and explanation without intent to infringe.

ISBN: 978-1-032-54626-1 (hbk)
ISBN: 978-1-032-55070-1 (pbk)
ISBN: 978-1-003-42888-6 (ebk)

DOI: 10.1201/9781003428886

Typeset in Times
by Deanta Global Publishing Services, Chennai, India

# Contents

# Preface

We take great pleasure in presenting this book, *Blockchain and IoT*, aimed at advancing knowledge and sharing recent applications and case studies in areas such as the Internet of Things (IoT), robotic and intelligent systems, web-based applications, and blockchain technology among working professionals and those engaged in education and research across various technical disciplines.

The objective of this book is to enable students to delve into IoT knowledge, which necessitates new security frameworks aligned with emerging technical standards. Consequently, the design of the book aims to address the need for new security paradigms for IoT based on these standards. Security, encompassing both physical devices and management applications, is paramount for the successful operation of IoT systems. The advantages of incorporating IoT in industrial automation include heightened efficiency, increased accuracy, cost-effectiveness, accelerated process completion, reduced power consumption, minimized errors, and enhanced control. Forward-thinking entrepreneurs aspire to automate their office processes to align with the latest technological advancements. However, it's important to acknowledge the challenges associated with automating industrial processes. In subsequent sections, we explore the potential synergy between blockchain and IoT technologies, investigating how they can complement each other effectively. The convergence of IoT and blockchain has the potential to reshape industries, redefining how we interact with technology, manage data, and conduct transactions. At its core, IoT empowers devices and sensors to communicate and exchange data seamlessly, creating a network of interconnected systems that gather real-time insights and drive efficiency. On the other hand, blockchain, renowned for its decentralized and immutable ledger technology, ensures transparency, security, and trust in data transactions, eliminating intermediaries and enabling peer-to-peer interactions. In recent years, the synergy between IoT and blockchain has gained significant attention, promising transformative solutions across various sectors, including healthcare, supply chain management, finance, energy, agriculture, and beyond. From optimizing supply chains and tracking assets to enabling secure micropayments and ensuring data integrity, the applications are diverse and profound.

This book serves as a comprehensive guide to understanding the dynamics, applications, and implications of integrating IoT and blockchain in industry. Through a blend of theoretical frameworks, practical case studies, and insightful analyses, we delve into the intricacies of this burgeoning field, exploring its potential to revolutionize business models, enhance operational efficiency, and drive innovation. Whether you are a seasoned professional, a curious enthusiast, or an industry leader navigating the digital transformation journey, this book provides valuable insights and practical guidance to harness the power of IoT and blockchain synergies. Join us as we explore the transformative potential of these technologies and embark on a journey toward a more connected, efficient, and transparent future.

# Editors

**R Anandan** completed his UG, PG, doctorate and post-doctoral degree (DSc) in computer science and engineering in Mexico. He is an IBMS/390 Mainframe professional and is recognized as a chartered engineer by the Institution of Engineers in India. He currently works as a professor, Department of CSE, School of Engineering; and is director of the Innovation and Incubation, Vels Institute of Science, Technology & Advanced Studies (VISTAS), Chennai, India. He has vast experience in corporate and all levels of academics in computer Science and engineering. He is currently supervising 8 PhD research scholars and 16 research scholars have been awarded PhD's under his guidance.

He is a member of many reputed international and national societies. He serves as an editorial board member, technical Committee member, or reviewer of several international journals. He has published more than 140 research papers in various international journals. He has presented 92 papers at various international conferences. He has received 18 awards from national and international agencies. He has authored and edited 28 books, including 10 international books, and published 30 chapters in various books. He has filed 18 patents of which 4 patents and 3 copyrights have been granted.

**D Akila** completed her post-doctoral fellow in computer science at Lincoln University College, Malaysia; and her PhD in computer science at Bharathiar University, Coimbatore, India. She received an MCA from Bharathiar University, MPhil from Bharathidasan University, and an ME in computer science and engineering from Vinayaka Mission University. She is currently a professor in the Department of Computer Applications, Saveetha College of Liberal Arts and Sciences, SIMATS Deemed University, Chennai, India. She has 15 years of teaching experience at both the undergraduate and postgraduate level. Her research interests include data mining, big data analytics, cloud computing, image processing, knowledge-based systems, and web information exploration. She has supervised two MPhil scholars and eight PhD scholars.

She is an active member of several professional bodies, including the IEEE, International Association of Engineers (IAENG) Hong Kong, Computer Science Teacher Association (CSTA) New York, Internet Society, International Society for Research and Development, and the World Academy of Science, Engineering and Technology (WASET). She serves as editorial board member, technical Committee member, or reviewer of many international journals. She has published more than 100 research papers in journals and has authored 7 Books.

She received a Life Time Achievement Award 2018–2019 for outstanding contribution in Education from the National IT Cyber Security Research Association, Government of India; the Achiever Award 2019 for outstanding contribution in the field of education from the Feminine Foundation; and from the Uyir Thalir Research and Resource Organization, A Excellence Teacher in Information Technology, Global Outreach Education Award 2019, New Delhi.

**Souvik Pal** is an professor in the Department of Computer Science and Engineering at Sister Nivedita University (Techno India Group), Kolkata, India. Prior to that, he was associated with Global Institute of Management and Technology; Brainware University, Kolkata; JIS College of Engineering, Nadia; Elitte College of Engineering, Kolkata; and Nalanda Institute of Technology, Bhubaneswar, India. Pal received his MTech and PhDin the field of computer science and engineering from KIIT University, Bhubaneswar, India. He has more than a decade of academic experience. He is the author or coeditor of more than 15 books from reputed publishers and holds three patents. He serves as a series editor for Advances in Learning Analytics for Intelligent Cloud-IoT Systems (Scopus-indexed); the Internet of Things: Data-Centric Intelligent Computing, Informatics, and Communication, published by CRC Press, Taylor & Francis Group, USA; and the Conference Proceedings Series on Intelligent Systems, Data Engineering, and Optimization, published by CRC Press, Taylor & Francis Group, USA. Pal has published a number of research papers listed in Scopus/SCI/SCIE journals and conferences. He was the organizing chair of RICE 2019, Vietnam; RICE 2020 Vietnam; and ICICIT 2019, Tunisia. He was invited as a keynote speaker at ICICCT 2019, Turkey, and ICTIDS 2019 and 2021 Malaysia. He also served as Proceedings Editor of ICICCT 2019, 2020; ICMMCS 2020, 2021; and ICWSNUCA 2021, India. His professional activities include roles as associate editor, guest editor, and editorial board member for more than 100 international journals and conferences . His research area includes cloud computing, big data, Internet of Things, wireless sensor networks, and data analytics. He is a member of many professional organizations, including MIEEE; MCSI; MCSTA/ACM, USA; MIAENG, Hong Kong; MIRED, USA; MACEEE, New Delhi; MIACSIT, Singapore; and MAASCIT, USA.

# Contributors

**B Chidhambararajan**
SRM Valliammai Engineering College
Kanchipuram, India

**Hariharan B**
SRM Institute of Science and Technology
Chennai, India

**Anwer Basha H**
Saveetha College of Liberal Arts and Sciences
Chennai, India

**Jeba Sonia J**
SRM Institute of Science & Technology
Chennai, India

**Maxraj J**
VELS Institute of Science, Technology and Advanced Studies
Chennai, India

**Edwin Jijo**
Military Technological College
Muscat, Sultanate of Oman

**Charanjeet K**
SRM Institute of Science & Technology
Chennai, India

**Revathi K P**
Easwari Engineering College
Chennai, India

**Balamurugan K**
SRM Madurai College for Engineering and Technology
Pottapalayam, India

**Priyadarsini K**
SRM Institute of Science & Technology
Chennai, India

**Senthil Kumar M**
SRM Valliammai Engineering College
Kattankulathur, India

**Azhagiri M**
SRM Institute of Science and Technology
Chennai, India

**Harsha Vardhini P A**
Vignan Institute of Technology and Science,
Telangana, India

**Jawahar P K**
B.S. Abdur Rahman Crescent Institute of Science and Technology
Chennai, India

**Latha Parthiban**
Pondicherry University Community College
Puducherry, India

**Anitha R**
B.S. Abdur Rahman Crescent Institute of Science and Technology
Chennai, India

**Parthiban R**
Sri Sivasubramaniya Nadar College of Engineering
Kalavakkam, India

**Selvakumarasamy**
SRM Institute of Science and Technology
Chennai, India

**Shenbagavadivu**
SRM Valliammai Engineering College
Chennai, India

**Anusooya S**
B.S. Abdur Rahman Crescent Institute of Science and Technology
Chennai, India

**Karthik S**
SRM Institute of Science and Technology
Chennai, India

**Mubeena S**
B.S. Abdur Rahman Crescent Institute of Science and Technology
Chennai, India

**Sandhya S**
SRM Valliammai Engineering College
Kattankulathur, India

**Susee S K**
SRM Madurai College for Engineering and Technology
Madurai, India

**Sahunthala S**
Rajalakshmi Institute of Technology
Chennai, India

**Selvakumarasamy Subramaniyam**
SRM Institute of Science and Technology
Nagar, India

**VijayAnanth Suyamburajan**
Vels Institute of Science, Technology and Advanced Studies
Chennai, India

**Manikandan T**
Rajalakshmi Engineering College
Chennai, India

**Jean Shilpa V**
B.S. Abdur Rahman Crescent Institute of Science and Technology
Chennai, India

# *Section–I*

## *Foundations in Block Chain*

# 1 Architecture, Security, and Privacy for Blockchain Using Enabling Technologies

*Priyadarsini K, Karthik S, R Anitha, Jean Shilpa V, Anusooya S, Jeba Sonia J, and Charanjeet K*

## INTRODUCTION

In 2009 Nakamoto Satoshi developed blockchain to make Bitcoins more secure. As of January 2018, the capital fee of Bitcoin was $180 billion. Blockchain is a multidisciplinary topic that involves mathematics, computation, big data, IoT, and cloud computing. Blockchain isn't always a product, however, it is an alternative as it is also used as the primary software of Bitcoin. A block is part of the blockchain that consists of a block header, which in turn is divided into fundamental block hash, block model, timestamp, nonce value, simple records, information, and Merkel root hash. The block also includes interactive records, input, and output. In general, it is an interconnected time series of information with cryptography and a dispensed shared ledger. It performs unusual operations by using consensus, smart contracts, preferred ledger, and cross-chain generation. Each block is connected to different blocks (nodes) and knows the previous and subsequent hashes in a peer-to-peer network. Blocks cannot be deleted or changed by blockchain users. It has a verification facility by which individual blocks have to be completely verified. Blocks have the traits of decentralization, persistency, anonymity, independence, openness, immutability, safety and agreement, collective renovation, programmability, transparency, unforgeability, and auditability [5] (Figure 1.1).

A Bitcoin transaction is a public ledger in the form of a list of instructions or information recorded in blocks as the Bitcoins are exchanged among users. Customers in transactions participate through Bitcoin addresses [16, 29] and they're controlled by using wallets. A user is not limited to one address; they can access several Bitcoin addresses in their wallets. These addresses are mapped to unique non-public/public key pairs through transformation characteristics [18]. A transaction among addresses, say, a and b, is expressed as follows:

DOI: 10.1201/9781003428886-2

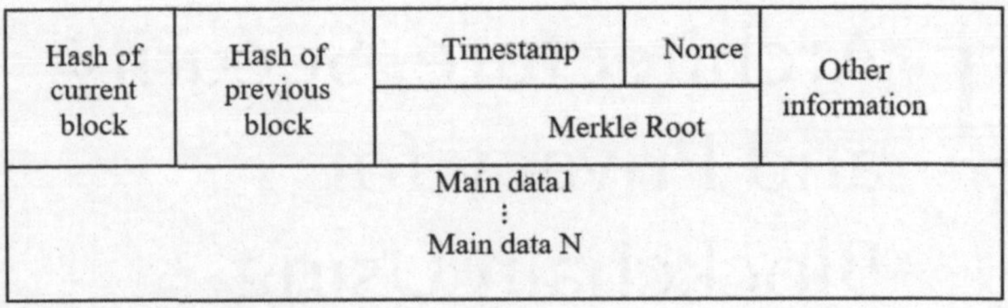


**FIGURE 1.1** Block structure.

(ab) = {source, BTC, b, *Sig* ska (source, BTC, b)}

where *Sig* ska is the signature with the aid of a private key and BTC is the Bitcoin amount to be transferred. There are a few shadow addresses that are mechanically created when transactions change values. A transaction CS (P, C, d, t, s) among P and C is carried out and there may be a CS. Devote, CS. Open, and CS. Nice stages denoting successful execution, starting, and aborting the transaction. If the transaction is completed successfully by secret key s, the Bitcoin price d is assigned to the receiver. There are several types of metadata: text, which uses many textual content messages; hash, a code that specifies the possession of files in the blockchain; financial document, a file of virtual payments or property; copyright, which allows manufacturers to promote and display the file; script, the many languages used for blockchain; photo, photos are sometimes also present in a blockchain; and archive, records saved in the compact form (Figure 1.2).

The authority is the centralized authority in a centralized device who has all the rights to update the database without needing third-party permission. In a blockchain, the advantage is that the consent of anybody might be required to perform an operation. It minimizes the monopoly of the relevant authority [24, 28, 46]. There are numerous varieties of blockchain: private for a private enterprise of small range; public or all, which has a wider range; and consortium for a group of organizations. They may be analyzed on the basis of the following parameters.

1. Only a few private nodes can perform the validation process of consensus determination.

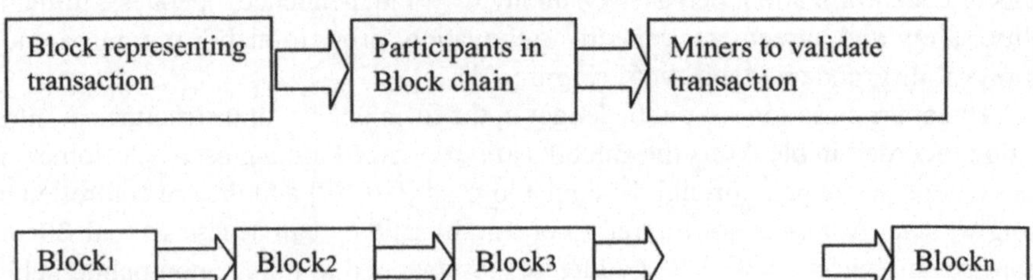


**FIGURE 1.2** Chain of blocks.

2. Only a consortium or organization has the power to decide whether information will be public or private.
3. Immutability: Blockchains are difficult for outsiders to tamper, while the consortium or organization is authorized to alter with the blocks.
4. Efficiency: The fewer the number of operations, the better the efficiency.
5. Centralized: It is controlled by one organization.
6. Consensus process: Only certified nodes are able to join the consensus process.

Many cryptocurrencies have been launched (Table 1.1). Some isolated cryptocurrencies, i.e., zero coin and zero cash based on zero knowledge proof, have been proposed [26, 30].

The articles selected for this chapter are from 2009 to 2021 (Figure 1.3).

## Architecture of Blockchain

The blockchain architecture is a collection of layers showing the functions and protocols supported by each layer [1, 2, 7, 32–34]:

i. **Data layer**: It creates the blocks and has its own information, verification, timestamp, block numbers, hash values of its predecessor, and successor nodes.
ii. **Network layer**: It makes a complete network of nodes via a peer-to-peer network with no fixed topology. Nodes can verify, transmit, and route for interactions.

**TABLE 1.1**
**Different Cryptocurrencies**

| Cryptocurrency | Year of Launch | Hashing | Mining |
|---|---|---|---|
| Bitcoin | 2008 | SHA-256 | Agreed PoW is done on all nonces |
| Litecoin | 2011 | S crypt | Same as Bitcoin PoW |
| Peercoin | 2012 | SHA-256d | PoW and proof of stake |
| Primecoin | 2012 | Cunningham Chain | PoW |
| Ripple | 2014 | Digital Signature | Consensus based |
| Ethereum | 2014 | Ethash | PoW |
| Permacoin | 2014 | Floating Digital Signature | Proof of retrievability |
| BlackCoin | 2014 | S crypt | Proof of stake |
| Auroracoin | 2014 | S crypt | PoW |
| Darkcoin | 2014 | X11 | PoW |
| Namecoin | 2015 | SHA-256d | PoW |

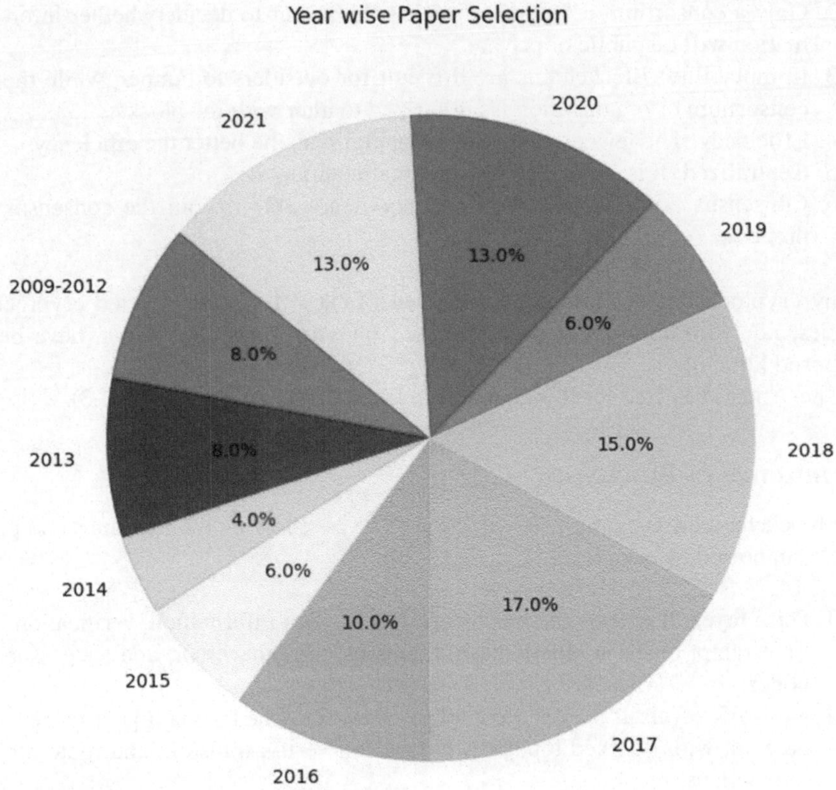


**FIGURE 1.3** Year-wise paper selection.

iii. **Consensus layer**: It is used to store accurate information by nodes in the shared ledger. It exhibits a higher satisfaction level in consensus efficiency. In a peer-to-peer network, all nodes work together to reach a consensus by Byzantine consensus. There could also be an oral or written consensus. After a block miner signs the blocks and broadcasts them, the receiver signs a new block after verification. The algorithm works on a majority of 51%. There are four types of consensus mechanisms: POW (proof of work) (Figure 1.4), proof of stake (POS), practical Byzantine fault tolerance (PBFT), and delegated proof of stake (DoPOS). Some other mechanisms are proof of authority (POA), proof of bandwidth, proof of elapsed time (POET).

iv. **Contract layer**: It is the lowest layer to make a blockchain operational via sophisticated programming. Examples are Ethereum and Definity. There are vulnerabilities in contracts, e.g., errors in encoding, cryptography, misaligned incentives, transaction ordering, timestamp dependence, mishandled exceptions, reentrancy, and data feed.

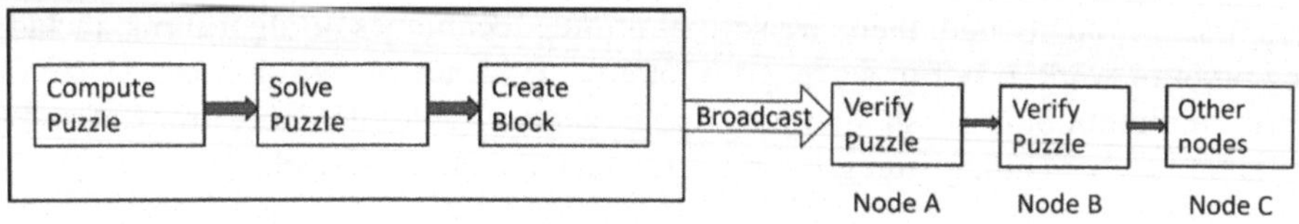


**FIGURE 1.4** PoW consensus mechanism.

v. **Service layer**: It provides the services to the upper layer by software and hardware.
vi. **Application layer**: It offers services and platforms to applications like IoT, smart city, and VANET (Figure 1.5).

Numerous styles of modeling are proposed for Bitcoin transactions. In the primarily graph-based approach, four types of transactions—namely, cash transfers, smart contract creation, smart contract invocation, and those not conducted through graph construction—are identfied and analyzed using graph evaluation techniques. The last transactions are labeled as a money drift graph, smart contract creation graph, and settlement invocation graph. The stochastic model is created to view the converting dynamics and behavior of the gadget. In the queuing version, each time nodes are introduced or mined they are queued up in a blockchain [23, 27].

## Blockchain Mining

Consensus or mining is needed to test the trustworthiness of nodes. These nodes can talk to different nodes without involving a third party. All nodes are linked to a

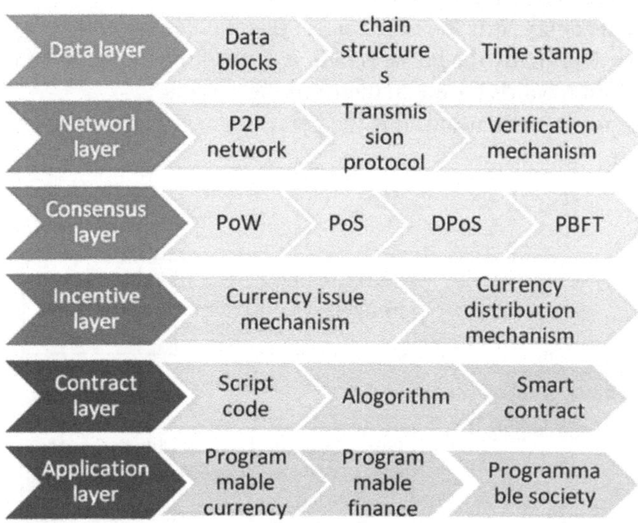


**FIGURE 1.5** Bitcoin architecture.

database. As mentioned, there are several mining techniques or algorithms, including proof of work, proof of stake, proof of area, evidence of significance, degree of trust, minimum block hash, and practical Byzantine fault tolerance. Proof of stake calls for less computing strength for mining than the proof of work.

## SELFISH MINING

Blocks are brought to the blockchain and mined via block miners. There are miners with their own mining powers, $mi$, such that $\sum I = n\ mi$. To check the validity of a block, hard crypto puzzles are solved by miners. Sometimes puzzles are too hard for one miner to solve, therefore, miners will pool together to collect enough computing strength. Miners get their rewards in Bitcoins once they solve the puzzle.

Extra computing energy will increase the rewards and the possibility of solving the crypto puzzles. Mining pools result in two types of companies: honest and egocentric [21, 35].

Egocentric miners convince the genuine miners to fix stale blocks with the aid of a forking blockchain, creating a copy of the blockchain (Figure 1.6). As a result, they waste their computing energy in futile efforts. Mining should be sincere and incentive-free. If a couple of block is introduced through miners, they create a fork into branches of a tree referred to as Merkel tree, which can be difficult or smooth depending on the basis of old and new regulations.

i. **Two types of nodes**: (1) Byzantine nodes, which behave arbitrarily or misbehave, and their mining strength is 25% less than the total computing power of all nodes and (2) honest nodes behave virtually. Consensus protocols are measured [15–22, 37, 45] by means of the following parameters to evaluate the performance and examine them.
ii. **Consensus delay**: It is the time that a system takes to reach an agreement.
iii. **Fairness**: Optimally the fairness ratio is 1, meaning the largest and smallest miners earn rewards based on their mining powers.
iv. **Mining power utilization**: It is the percentage of the power used to secure the mining power.

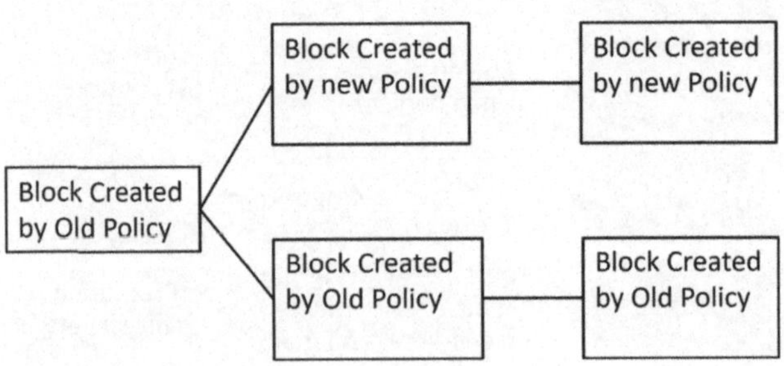


**FIGURE 1.6** Forking of blockchain.

**TABLE 1.2**
**Analysis of Consensus Algorithms**

| Consensus Protocol | Performance | Need of Token | Maximum Malicious Nodes | Type | Attacks | Special Hardware |
|---|---|---|---|---|---|---|
| PoW | High | Yes | 51% | Public | Yes | No |
| PoS | High | Yes | 51% | Public | Yes | No |
| DPoS | High | Yes | 51% | Public | Yes | No |
| PBFT | High | No | 33% | Consortium | No | No |

v. **Subjective time to prune**: Because of its probabilistic nature, Nakamoto consensus is not sufficient to identify the pruned branches in Bitcoins.
vi. **Time to win**: The ratio of the first time to assume that node is never pruned to when that node was pruned.
vii. **Node identity**: It is an identity that is needed to select a primary miner.
viii. **Energy saving**: Hashing requires energy, so algorithms that do not use hashing can save the energy, such as the ripple miner.
ix. **Tolerated power of adversaries**: It is 51% of the hash power.
x. **Efficiency**: It is measured by the propagation of time and inactivity. Both are desired to be minimum.
xi. **Centralized**: Public blockchains are decentralized, consortiums are semi-centralized, and private blockchains are fully centralized.
xii. **Consensus process**: In a public blockchain, anybody can participate. In a consortium, a certificate is needed to participate in a consensus process (Table 1.2).

### Bitcoin Wallets

There are many Bitcoin wallets, including TREZOR, MultiBit Core, Coinbase, Blockchain, Exodus, MyCelium, Bitcoin middle, MultiBit HD, Electrum, Copay, Amory. For security and privacy, Coinbase, MyCelium, Bitcoin middle Blockchain, TREZOR, Exodus, MultiBit middle, MultiBit HD are desirable, whereas Electrum, Copay, and Amory are not. They're unbiased wallets, except Coinbase. In large experience, Bitcoin wallets are classified into types 0, 1, and 2, on the basis of the capacities of logistic assets they require [31] (Figure 1.7).

## Key Enabling Blockchain Applications

The proposed technologies fintech, IoT, cloud, governance, and corda have helped in the development of Bitcoin [2, 10, 32].

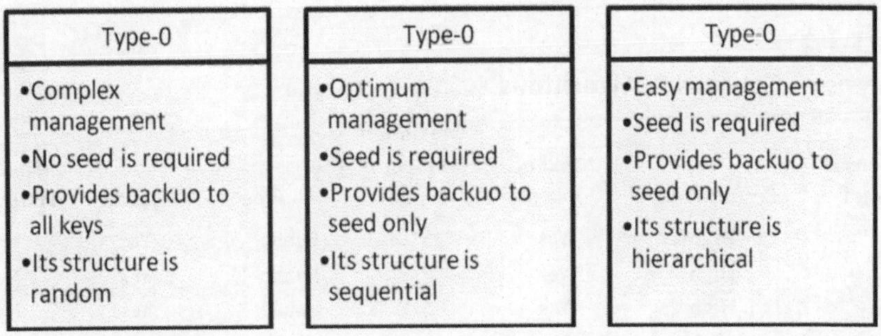


**FIGURE 1.7** Wallet classification.

### Fintech

Since operations of Bitcoins are unforgeable, accurate, and untraceable, through the smart contract agreement with the regulations, blockchain has increased the confidence of the financial industry. It provides security to the upper layer and multi-signature scheme messages are encrypted to improve the verification and anonymity.

### Internet of Things (IoT)

For high protection, a blockchain can utilize the IoT infrastructure. There are challenges in IoTs: system design (centralization bottleneck, lightweight design), facts management (processing tool renovation service management payment and discovery services), and protection (privacy, counterfeit, authentication, secure communique, and counterfeit hardware). Blockchain usage in IoT involves too much overhead and scalability issues. Consequently to reduce the overhead and to beautify the authentication and authorization of factors, a three-layered modeling approach is proposed in [42–44]. In layer 1, there may be a cluster head of IoT devices and security offerings presented via local cluster heads. The cluster head is supposed to perform registration of the latest node, consultation key and conversation control, and secure communique. In the second layer, cluster heads function under a base station and they gather and forward the data to the higher layer. Layer 3 collects the bottom stations to prepare the owner to control the tool to supply the statistics in the cloud server. IoT answers the challenges of hardware, records analytics, maintenance, mobility, safety, and logistics. IoT to IoT peer, IoT peer to blockchain, and hybrid architectures are proposed. There are many problems to the mixing of blockchain with artificial intelligence for sharing, securing, transactions, deposit, and aid control programs.

### Supply and Chain Management

It has many events of logistics, capital flow, and information flow; there must be collaboration between entities to keep supply intact via blockchain. Various encryption and decryption e.g. symmetric and asymmetric, DES, AES, RC, RSA, Elliptic curve, Diffie-Hellman, MD, SHA, Merkle tree, etc.

### Cloud

A cloud is a network of blocks and a database of nodes of a blockchain is considered big data. Therefore, a blockchain would require a cloud to store and manage the database because of its size, volume, veracity, and velocity.

### Governance

It is desired to increase the digital transmission of financial institutions, digitization of important government records, land registries, patients' records, and so on. Currently, flexible regulations may need to be standardized and regulated to cope with attacks, threats, and vulnerabilities.

### Corda

Corda is a public digital ledger technology, not a blockchain, because it creates a network by which messages and other information flow. It is more reliable, transparent, immutable, reduces risk, flexible, and regulated. It controls financial transactions by recording financial agreements, validating transactions between two parties, and providing the perfect flow of information in a regulated and supervised manner. Corda automatically produces legal document models representing individuals' legal rights. Corda uses PoA, is permissioned, and is not suitable for use-case. A drawback is the generation of millions of long-lasting threads [38]. Permissions of the blockchain are not like the regular ones because it is not more secure. It can connect to any node, which can add blocks to the blockchain.

- **Applications**

In the economic sense, traditional Bitcoins need built-in incentives or rewards, a centralized and decentralized Bitcoin economic system, currency exchanges, mixers, digital wallet services, and mining pools (Figure 1.8 and Figure 1.9). There are intermediaries that helped in the development of Bitcoin [14]:

i. **Currency exchanges**: Like most traditional currencies, it charges from 0.2% to 2%.
ii. **Digital wallet services**: Wallets contain account information, private/public information, and recorded transactions. Some users use private wallets that offer a wide range of services. The files are shared through web servers or mobiles.
iii. **Mixers**: The transactions are mixed in random order so as to prevent any guess or tracking of transactions.
iv. **Mining Pools**: Bitcoins are permitted by the miners after solving the puzzle (need high computational power). So instead of having only one miner, miners are kept in a row to speed up the process.

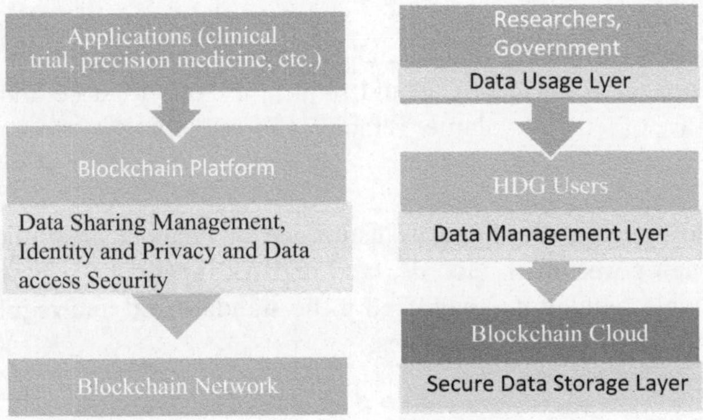


**FIGURE 1.8** Layers applicable to traditional blockchain.

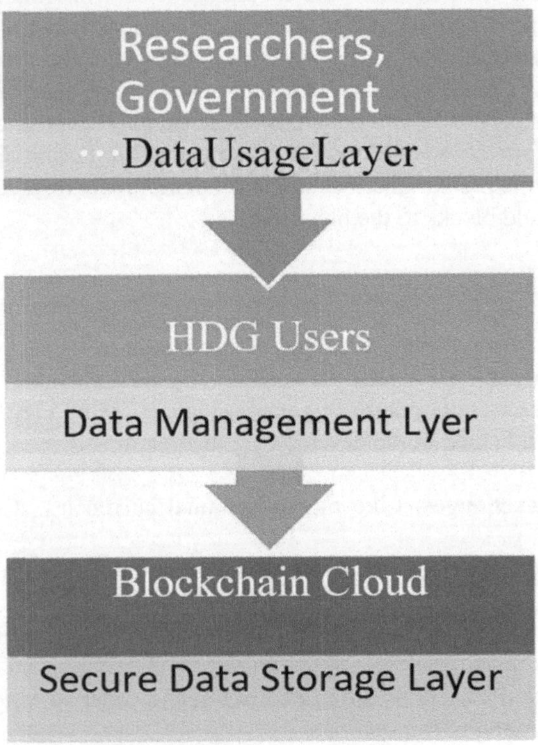


**FIGURE 1.9** Layers in data management in blockchain.

## Open Source Blockchain Implementation

Bitcoin miners need cryptographic strategies, consensus methods, smart contracts, and peer-to-peer networks. There are diverse styles of proof strategies, including PoW, PoS, PBFT, and DpoS. The cryptography virtual signature algorithm is used

**TABLE 1.3**
**Open-Source Bitcoin Implementation**

| | |
|---|---|
| Chain | Interface for total blockchain nodes to know activities done by address, i.e., balance, notifications to node where queries are made by address. |
| Stellar | It is an API development environment for developing semi-centralized public ledger. Block.io, Gem, and block ciphers are associated to this. |
| Block cipher | It contains APIs of a multi-signature wallet. |
| IPFS project | It is a global file system that suggests decentralize secure file serving and sharing. BitTorrent is a peer-to-peer file sharing. It is similar to Filecoin. |
| Internet archive | To store the blockchain along with ledger. |

for authentication. For public key encryption gadgets there are RSA, Diffie–Hellman, and elliptic curve cryptography. A collection of nodes without any centralized controller in which each node acts as a router, verifier, or miner of the blocks and wallets is called a peer-to-peer community. All nodes in a collaborative community are created on a public ledger and are visible and verified by way of all nodes inside the community. Smart contracts are better than the traditional ones considering the feel of pace and guidelines. Transactions continue executing until some predicted consequences are determined. When no expected outcomes are visible in the allotted time, the transaction is aborted [39]. The open-source implementations Bitcoin, Ethereum, and Hyperledger are described next [1, 10, 11, 25] (Table 3.1 and Figure 1.10). Xblock-ETH, Xblock-EOS, Bockci, Blockbench, NodeFinedr, and network simulator facts set framework of Blockchain are described in the literature [40].

i. **Bitcoin**: It uses POW mining and no smart contract. They are more popular and good for scalability for a number of nodes, but it is expensive computationally and time consuming.
ii. **Ethereum**: It uses the smart contract, POW, and POS mining. It is scalable and it does not require any computation for mining, but it is costly.
iii. **Hyperledger**: It requires a smart contract and PBFT, and it is the fastest to detect the validity of the transactions. It records transactions and it is accessible to all nodes. It cannot be added, deleted, or modified, meaning it is fully protected from attacks.
iv. **Blockchain APIs and development platforms**: Chain.info consists of APIs with wallets to perform payments and other operations [13].
v. **Multichain, Hydrachain, Bigchain, Openchain, Quorm, Chaincore**: They are privately permissioned and they are developed in Java and Python. They are used in financial operations, banking, land registries, and ID verification (Table 1.3).

Bitcoin transaction contain a version number, in-counter for the number of inputs, out-counter for the number of outputs, previous transaction, previous out-index,

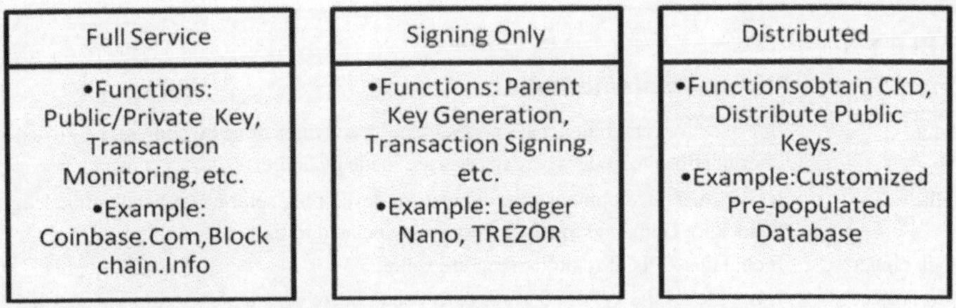


**FIGURE 1.10** Open-source Bitcoin services.

**TABLE 1.4**
**Scripting Commands**

| **OP_CHECKSIG** | **Opcode** |
|---|---|
| *Sig* | Signature |
| in-script:*sig*<br>out-script:*pubKey*OP_ CHECKSIG | Pay to public key: this is used to verify the signature by pushing *sig* on top of stack, then pushing *pub Key*, and finally OP_CHECKSIG. |
| in-script: *sigpubKey*<br>out-script: OP_DUP_ HASH 160<br>*pubKeyHash*<br>OP_ EQUQLVERIFY OP_<br>CHECKSIG | Pay to public key hash: This is same as above except here double hashing of public key is used. |
| in-script:OP_0*sig*1···*sigM*<br>out-script:*MpubKey1*···*pubKeyNN*<br>OP_CHECKMULTISIG | Multi-Signature: It does the multi-signature verification. |
| in-script: *v1*···*vN*bitstring<br>out-script: OP_HASH160 hashOP_EQUAL | Pay to script hash: It is used to make sure that bitstring is equal to hash. |
| out-script: OP_RETURN bit string | It removes false values from the set. |

in-script, sequence numbers, the value to be transferred, out-script length, out-script, and lock time to lock the transaction. A stack-based language script has a limited cryptographic and logical operations for embedding data in transactions [36] (Table 1.4).

## Issues and Challenges

Since Bitcoin is a growing technology, it has limitations, including technical, business model, government regulation, and privacy challenges for personal data. Blockchains have the following issues and challenges. Bitcoin size is limited to 1 MB. It is mined every ten minutes and can handle seven transactions per second; therefore, it requires a larger storage and lower propagation delay. It has no ownership, which results in a management issue. Other challenges are resource sharing including energy, authentication, verification, cross-lock chain, and decentralized storage. Some of the crucial challenges are as follows:

i. **Throughput or scalability**: A maximum seven transactions per second is possible while some systems like VISA and Twitter allow 2000 to 5000 transactions per second, respectively [12, 41]. Efforts to increase scalability include storage optimization and redesigning the blockchain.
ii. **Bandwidth, computing power, and size**: The size of blockchain is increasing year by year, so to process it requires more bandwidth and size. Electricity current consumption is 2.55 GW and the average electricity consumption per transaction is 300 KWH and more power is expected for the future.
iii. **Latency**: To complete a transaction takes around ten minutes. Contemporary systems like VISA take only a few seconds.
iv. **Resource wastage**: There a lot of resource wastage because of the proof-of-work process.
v. **Privacy leakage**: Though it is quite safe, many people can still generate many addresses due to information leakage. Bitcoins can be leaked to reveal their information. Therefore, anonymity methods are proposed like mixing. A small blockchain created by a small network has the 51% probability of being attacked [6].
   - Blockchains are analyzed by a transaction graph T (T, L), where T is the transactions set and L is the input–output relations.
   - Address graph A (A, L'), where A is the address of Bitcoins and L' is the directed assignments of connecting addresses.
   - Entity graph: It merges the addresses of the same users and it is represented by Ɛ (E, L"), where E is the set of entities.
   - Ownership: Mapping of addresses to destinations. This causes the analysis of using multiple addresses by the same user for multi-entry transactions; transactions with change, curious, or malicious mixing services; blockchain peer-to-peer network privacy; and web payments.
vi. **Selfish mining**: Some nodes do not broadcast their share and cheat, even if they have less than 51% hashing power for transmission.
vii. **Usage**: The development of API is challenging. An easy-to-use Bitcoin API is needed.

## Privacy, Security, and Risk

Bitcoin supports consistency, integrity, availability, tampering resistance, resistance to double-spending, pseudonymity, consensus algorithms, and signature verification public key as pseudonyms [48–52]. They are supposed to incorporate unlinkability, confidentiality, resistance to majority consensus attack mixing, and anonymous signatures. With Bitcoin privacy quantification [1, 17], an adversary is not allowed to link two addresses or two transactions of a user, known as activity unlinkability.

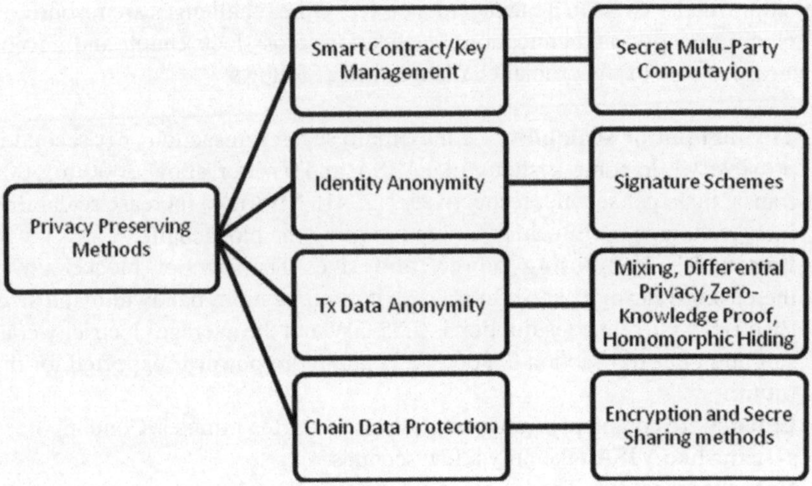


**FIGURE 1.11** Privacy preserving terminology.

Profile indistinguishability refers to the inability of adversaries to reconstruct a user's profile (set of addresses or set of transactions of Bitcoin users) in a public ledger (Figure 1.11).

### Techniques to Enhance the Privacy and Anonymity of Bitcoins

i. **Coinbase**: Transactions require multi-signatures. Anonymity level varies.
ii. **Coin shuffle**: It is used for coordinating Coinbase transactions via cryptographic protocols and lower anonymity levels.
iii. **Xim**: Multiround partnering mixing at a higher mixing rate.
iv. **Coin Shuffle and DiceMix**: Optimal mixing is done to improve the anonymity of cryptocurrency and DdoS may occur. It does not use a third party and less risky, but it is less resistant to DoS attacks.
v. **Value Shuffle**: Coin Shuffle-based to achieve unlinkability and better privacy.
vi. **Dandelion**: Policy to avoid the deanonymization of Bitcoin.
vii. **Secure Coin**: It is secure for anonymization and unlinking.
viii. **CoinParty**: It helps to understand the advantages of centralized and decentralized mixing, and also suspected coin theft.
ix. **Mix Coin**: Mixing is done by a third party. Partial theft resistance and unlinkability.
x. **Blind Coin**: Blind signatures are used for anonymity and provide partial theft protection.
xi. **Tumble Bit**: Payments conducted via untrusted intermediaries and require double back-to-back transactions.

### Security

Security means providing the access control, confidentiality, integrity, authentication, availability, and non-repudiation of services. Miners whose computing power is below 25% of the network's total power are incentivized. Therefore, they are lured to use small blocks and extend the longest and heaviest chains. The honest users follow the heaviest chain while shellfish users may follow any chain. To increase their revenue, miners prefer to follow the longest blockchain fork. Bitcoins are more flexible in network optimal conditions, stating zero attacks in zero latency. Secure multiparty computation, attribute-based mechanisms, and non-interactive Bitcoins are more flexible in zero-knowledge proof. Blockchain provides very good authentication for access control with the help of hashing [47]. Some of the security issues are as follows:

- **Wallet security**: Wallets need to be protected from the attacks to get the private key.
- **Censorship resistance**: To limit the power of the lead miner to a single epoch.
- **Resilience to mining power variation**: They are adjustable to different mining power variations.
- **Forks**: Making of other blockchains, which wastes the efforts of honest nodes.
- **Double-spending**: Nodes try to transfer the same Bitcoin amount to more than one node.

### Classification of Attacks

Attacks compromise the security of the system. Attacks are classified into several different categories depending on where they pose the threats [4, 20]: identity based (replay, impersonation, and Sybil), manipulation (Eclipse, transaction malleability, time jacking, quantum, reputation), vulnerability in service (race, DdoS, double-spending, finney, vector76, collusion; malware), application vulnerability (use case design flow, coding vector) [8].

A. Bitcoin and proof of work attacks based on consensus protocol:

i. **Selfish mining attack**: Cheating to increase rewards.
ii. **DAO attack**: It is exploited by re-reentering the contaminated smart contracts.
iii. **BGP hijacking attack**: Intercepts the miners and reroutes the cryptocurrency so it can be stolen.
iv. **Eclipse attack**: It is used to separate the victim from the rest of the network.
v. **Balance attack**: It disrupts the communication of low-power miners.
vi. **Double-spending or race attack**: The seller or merchant is targeted, and the same Bitcoins are sent in rapid succession. They are countered by putting observers in between them.

vii. **Finney attack**: Sellers and merchants are affected. A faulty miner sends a pre-mined block for double-spending. It can be countered by multiple confirmations.
viii. **Brute force attack**: To secretly mine a blockchain and it is controlled by using the observers.
ix. **Vector 76 or one confirmation attack**: It is a fusion of double-spending and finney attacks and again it can be avoided by using an observer.
x. **More than 50% hash power or gold finger attack**: It manages adversaries with up to more than 50% hash rate and it can also be controlled by inserting observers.
xi. **Block discarding or selfish mining**: It misuses the fork feature to unfairly gain rewards. It can be controlled by the zero block technique and timestamp base mechanisms.
xii. **Block withholding**: It mines only PpoW. It can be avoided by involving only known and trusted miners.
xiii. **Fork and withholding attack**: It corrects the bad effects of block withholding and selfish mining. It is a node fence, therefore it is more dangerous.

B. Misbehavior attacks targeting Bitcoins are as follows:

i. **Bribery attacks**: Some nodes bribe to get advantages for mining, and they affect the seller and merchant. It can be countered by increasing the reward of honest miners.
ii. **Refund attack**: It is a refund method that is exploited by adversaries for payments, and they can be managed by publicly verifiable evidence.
iii. **Punitive and feather attacks**: A miner blacklists transactions of other miners. They affect the users and are difficult to control.
iv. **Transaction malleability**: An attacker changes the ID of a transaction.
v. **Time jacking**: An attacker speeds up the miner's clock. It can be prevented by restricting acceptance time ranges.
vi. **DdoS**: An attack in groups to exhaust the network resources, and it can be countered by authorization protocols.
vii. **Sybil**: Multiple identities are used by the attacker. It can be countered by mutual mixing protocols.
viii. **Eclipse or net split**: The attacker monopolizes network connections. Here attackers try to isolate the target from the network. It can be controlled by disabling the connections.
ix. **Tampering**: A delay is caused in transactions. It is managed by improving the block request management.
x. **Routing attacks**: Some nodes are skipped from the network. It is controlled by increasing the diversity connections.
xi. **Deanonymization**: Wallets are linked with IP addresses. It affects users only. It can be countered by mixing the Coin Join and Coin Shuffle services.

### Solutions to Bitcoin Attacks

These are some solutions to prevent the attacks:

i. **Double-spending attack**: Confirmations are received from observers. The methods exist like listening period, inserting observers, and forwarding double-spending attempts. For detecting the attacks, non-equivocation contracts are employed to lock the Bitcoins in deposit with the payee. When a payee double-spends, cryptographic methods reveal it and penalize the malicious payer.
ii. **Forking and pooling**: This is a backward compatible means when a miner gets many forks of the same length, and will forward them to all other peers to select any one fork randomly. Other techniques use timestamps to check the recency.
iii. **Securing wallets**: A wallet contains the private and public keys for each account, which is encrypted by AES, CBC, SHA, elliptic curve digital signature methods, etc. Another wallet called a cold wallet is kept unconnected to the network as a standby, and some Bitcoins are transferred to this cold wallet. Therefore, the bank cannot access this wallet.
iv. **Securing Bitcoin networks**: It helps to secure networks from DdoS attacks, refund attacks, and transmission malleability.
v. **Defense against anonymity, peer-to-peer network**: Certificateless authentication, convergence encryption, counterresistance to attacks, and blockchain-based distributed key management architecture are proposed to be employed.

### Risks to Blockchains

Various dangers or threats to Bitcoins are considered here [3] (Table 1.5). They are vulnerable to 51% attacks. It results in many attacks such as opposite transactions, reverse ordering, hampering of mining transactions, and confirmation of ordinary transactions. Private key security, referred to as identification, is dealt with by a third party. Some criminal activities can be managed like ransomware, underground marketplace, money laundering, and double-spending. Transactions are traceable and their order of execution needs to be recorded in a defensive storage. Transaction privacy is not secure and criminals can conduct malicious activities through smart contracts risking information leakage. The fee of operations is judged by the execution time, bandwidth, computing, and resources. In under-priced operations, the attacker launches attacks that consume a lot of resources. Marketplace risk is due to the versions inside the foreign exchange quotes. In shallow marketplace problems, a few parties hesitate to obtain or pay the high amount because of the low quantity of Bitcoins.

## Sustainability and Potential Failures

Foreign money is safe from the authorities' regulations for business and purchasers. However, still there are some gray areas concerning restrictions using Bitcoin because of its decentralization coverage and technical failures. Some countries have regulated Bitcoin by way of recognizing or no longer backing it. Many federal governments have a monopoly on issuing currencies. A few nations do let states issue

**TABLE 1.5**
**Bitcoin Risks**

| Risk | Cause | Range of Influence |
|---|---|---|
| 51% vulnerability | Consensus mechanism | Blockchain 1.0, 2.0 |
| Private key security | Public-key encryption | |
| Criminal activity | Cryptocurrency application | |
| Double-spending | Transaction verification method | |
| Transaction privacy leakage | Transaction design flaw | |
| Criminal smart contracts | Smart contract application | Blockchain 2.0 |
| Vulnerabilities in smart contract | Program design flaw | |
| Under-optimized smart contract | Program writing flaw | |
| Under-priced operations | EVM design flaw | |

local currencies. Some currencies are recognized with the aid of governments. This section analyses some concerns surrounding Bitcoin [19]:

i. **Iraqi Swiss Dinar**: This currency was neither recognized by the government nor a commodity, but never collapsed.
ii. **Confidence**: Many people are afraid of investing in Bitcoins because of technical and legal issues like improper use of discretionary authority, superior and competent currency, government crackdown, and deflationary spiral.
iii. **Anonymity failure authentication problem**: Since transactions are in public and online, there can be account identification issues and breaking of the anonymity. It will degrade the performance of Bitcoins as well as the authenticity of the blockchain.
iv. **Theft**: Since it is in the form of digital currency, it is possible that it can be lost or stolen.
v. **Denial of service**: Some malicious nodes have enormous computing power and can disrupt transaction processing.
vi. **Integrated cost and hesitancy**: To maintain a blockchain, a lot of computing resources are required. Because of some legal and economic risks, there exists some hesitance in using blockchain publicly for public transactions.
vii. **Fraud**: Many consumers want fraud protection. Bitcoins are stolen because they are digital virtual cryptocurrencies.

## Blockchain Applications

Some of the main applications of blockchains are as follows (also see Figure 1.12):

i. **Finance**: Financial services, P2P financial market, risk management, enterprise transformation, financial transactions.

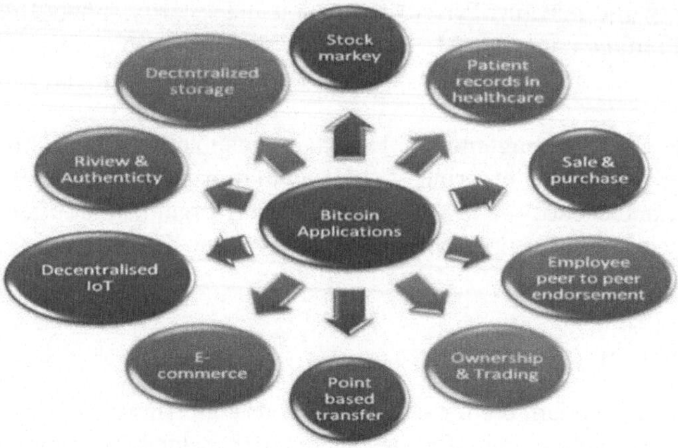


**FIGURE 1.12** Some Bitcoin applications.

ii. **Internet of Things**: Safety and privacy in applications [9].
iii. **Public and social services**: It is useful in land registration, education, insurance, freedom of expression, and energy where parties can share identities and control data privately and securely.
iv. **Health**: Doctors use the validated correct block appended to the blockchain to preserve the privacy of the patient. Health professionals share electronic health records (HERs) to diagnose and treat patients. Some measures such as access control should be provided since health data is highly private. There should be the option to automatically remove data after certain uses or time. The size of healthcare data is also very large and needs separate off-chain storage strategies. For these issues, a Healthcare Gateway (HDG) is proposed. Table 1.6 lists blockchain uses in healthcare.
v. **Reputation system**: Measures trustworthiness of users.
vi. **Socio-technical security**: This perspective acknowledges that effective blockchain security relies not only on cryptographic protocols and technical safeguards but also on understanding human behavior, user interactions, governance structures, and the socio-economic context in which the technology operates.

**TABLE 1.6**
**Blockchain in Healthcare**

| Study | Year of Launch | Hashing | Mining |
|---|---|---|---|
| HDG | 2016 | Nothing | Nothing |
| MedRec | 2016 | Ethash | PoW |
| PSN | 2016 | Nothing | Nothing |
| BBDS | 2017 | SHA-256 | PoW |

vii. **Security and privacy**: Privacy protection and security enhancement.

viii. **Data management and AI**: Bitcoins are useful to manage big data on IoT and they are also helpful in applications involving artificial intelligence (AI).

ix. **Supply chain management**: Blockchain technology manages the distributed ledger, and the ordering of goods with transparency, reliability, traceability, and security. The supply chain of any organization often involves the raw material supply, wholesaler, manufacturer, distributor, retailers, and end users.

## RESULTS AND DISCUSSIONS

Blockchain is the modern era for safe transactions concerning finance, privacy, IoT, agreements, and public social offerings. Its features include a decentralized mechanism to validate the recorded transactions within the public ledger. Its protocols are evolving, however, they need a lot of security and privacy. In many nations, they are free from government oversight. There are a few flaws that breach the security of Bitcoins. This chapter describes blockchain generation, blockchain structure, consensus or mining mechanisms and their evaluation, problems and challenges, assaults, and packages. Protocols are very well described, compared, and analyzed (Figure 1.13).

## CONCLUSIONS

Blockchain is an integration of information technology, IoT, and machine-to-machine (M2M) interactions. It has scope for digital artificial intelligence (smart contracts), IoT, and consensus mechanisms based on collaboration for resolutions. There are unique elements, for example, safety and privacy, threats, attacks, and their implementation through open sources. Bitcoin is useful because of its security,

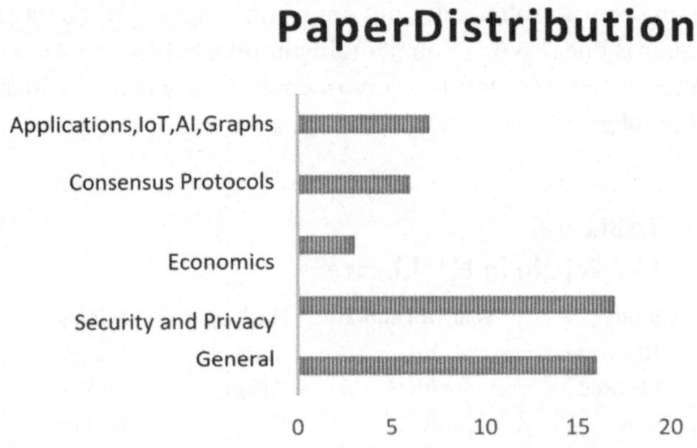


**FIGURE 1.13** Subject-wise paper distribution.

trust, public validity checking, and its restricted guidelines. It's one of the safest and maximally secure foreign money in the arena. Worldwide, people are preferring shopping for and doing commercial enterprise in cryptocurrencies. This chapter covered its safety, methodologies, processing, designing, mining, and validating mechanisms. It also reviewed the developments regarding key control (privateness and protection), identity control, solving difficult crypto puzzles, caching of records, and mobility.

## REFERENCES


1. Salman T, Zolanvari M, Erbad A, Jain R, Samaka M. Security services using blockchains: A state of the art survey. *IEEE Communications Surveys & Tutorials*. 2018; 21 (1):858–880. https://ieeexplore.ieee.org/document/8428402.
2. Lu Y. Blockchain: A survey on functions, applications and open issues. *Journal of Industrial Integration and Management*. 2018; 3 (04):1850015. https://www.worldscientific.com/doi/10.1142/S242486221850015X.
3. Zheng Z, Xie S, Dai HN, Chen X, Wang H. Blockchain challenges and opportunities: A survey. *International Journal of Web and Grid Services*. 2018; 14 (4):352–375. https://www.researchgate.net/publication/328338366_Blockchain_challenges_and_opportunities_A_survey.
4. Li X, Jiang P, Chen T, Luo X, Wen Q. A survey on the security of blockchain systems. *Future Generation Computer Systems*. 2020;107:841–853.https://doi.org/10.1016/j.future.2017.08.020.
5. Herlihy M. Blockchains from a distributed computing perspective. *Communications of the ACM*. 2019; 62 (2):78–85. https://dl.acm.org/doi/10.1145/3209623.
6. Tschorsch F, Scheuermann B. Bitcoin and beyond: A technical survey on decentralized digital currencies. *IEEE Communications Surveys & Tutorials*. 2016; 18 (3):2084–2123. https://ieeexplore.ieee.org/document/7423672.
7. Dinh TT, Wang J, Chen G, Liu R, Ooi BC, Tan KL. Blockbench: A framework for analyzing private blockchains. In *Proceedings of the 2017 ACM International Conference on Management of Data 2017 May 9*, 2017, pp. 1085–1100. https://arxiv.org/abs/1703.04057.
8. Bhushan B, Sinha P, Sagayam KM, Andrew J. Untangling blockchain technology: A survey on state of the art, security threats, privacy services, applications and future research directions. *Computers & Electrical Engineering*. 2021;90:106897. https://doi.org/10.1016/j.compeleceng.2020.106897.
9. Ali MS, Vecchio M, Pincheira M, Dolui K, Antonelli F, Rehmani MH. Applications of blockchains in the Internet of Things: A comprehensive survey. *IEEE Communications Surveys & Tutorials*. 2018; 21 (2):1676–1717. https://ieeexplore.ieee.org/abstract/document/8580364/
10. Sankar SL, Sindhu M, Sethumadhavan V. Survey of consensus on blockchain applications. *Int'l Conference on Advanced Computing and Communication System*, January 2017, pp. 1–4. https://www.researchgate.net/publication/319277442_Survey_of_consensus_protocols_on_blockchain_applications.
11. Alhadharmi, Z., Alghfeli, S, Alghfeli M. Zawish M. Introducing blockchains for healthcare. *Int'l, Conference on Electrical and Computing Technologies and Applications (ICECTA)*, 2017, pp. 1–4.
12. Yli-Huumo J, Ko D, Choi S, Park S, Smolander K. Where is current research on blockchain technology?—A systematic review. *PloS ONE*. 2016; 11 (10):e0163477. https://doi.org/10.1371/journal.pone.0163477.

13. Swan M. *Blockchain: Blueprint for a new economy.* O'Reilly Media, Inc.; 2015. https://dl.acm.org/doi/10.5555/3006358
14. Böhme R, Christin N, Edelman B, Moore T. Bitcoin: Economics, technology, and governance. *Journal of Economic Perspectives.* 2015; 29 (2):213–238. https://www.aeaweb.org/articles?id=10.1257/jep.29.2.213.
15. Eyal I, Gencer AE, Sirer EG, Van Renesse R. Bitcoin-NG: A scalable blockchain protocol. In *13th {USENIX} symposium on networked systems design and implementation ({NSDI} 16)*, 2016, pp. 45–59. https://www.usenix.org/conference/nsdi16/technical-sessions/presentation/eyal.
16. Brito J, Castillo A. *Bitcoin: A primer for policymakers.* Mercatus Center at George Mason University; 2013. https://www.mercatus.org/publications/technology-and-innovation/bitcoin-primer-policymakers.
17. Bentov I, Kumaresan R. How to use bitcoin to design fair protocols. *Annual Cryptology Conference 2014 Aug 17*, pp. 421–439. Springer, Berlin, Heidelberg. https://www.iacr.org/archive/crypto2014/86160326/86160326.pdf.
18. Bariviera AF. The inefficiency of Bitcoin revisited: A dynamic approach. *Economics Letters.* 2017; 161:1–4. https://www.sciencedirect.com/science/article/abs/pii/S0165176517303804
19. Grinberg R. Bitcoin: An innovative alternative digital currency. *Hastings Science & Technology.* 2012; 4:159. https://www.researchgate.net/publication/228199328_Bitcoin_An_Innovative_Alternative_Digital_Currency.
20. Conti M, Kumar ES, Lal C, Ruj S. A survey on security and privacy issues of bitcoin. *IEEE Communications Surveys & Tutorials.* 2018; 20(4):3416–3452. https://ieeexplore.ieee.org/document/8369416
21. Eyal I, Sirer EG. Majority is not enough: Bitcoin mining is vulnerable. *International Conference on Financial Cryptography and Data Security 2014 Mar 3*, pp. 436–454. Springer, Berlin, Heidelberg. https://www.researchgate.net/publication/258224002_Majority_Is_Not_Enough_Bitcoin_Mining_Is_Vulnerable.
22. Garay J, Kiayias A, Leonardos N. The bitcoin backbone protocol: Analysis and applications. *Annual International Conference on the Theory and Applications of Cryptographic Techniques 2015 Apr 26*, pp. 281–310. Springer, Berlin, Heidelberg. https://eprint.iacr.org/2014/765.pdf.
23. Ron D, Shamir A. Quantitative analysis of the full bitcoin transaction graph. *International Conference on Financial Cryptography and Data Security 2013 Apr 1*, pp. 6–24. Springer, Berlin, Heidelberg. https://eprint.iacr.org/2012/584.pdf.
24. Barber S, Boyen X, Shi E, Uzun E. Bitter to better—how to make bitcoin a better currency. *International Conference on Financial Cryptography and data security 2012 Feb.*
25. Antonopoulos AM. *Mastering Bitcoin: Programming the open blockchain.* O'Reilly Media, Inc.; 2017. https://github.com/bitcoinbook/bitcoinbook.
26. Lischke M, Fabian B. Analyzing the bitcoin network: The first four years. *Future Internet.* 2016;8 (1):7. https://www.mdpi.com/1999-5903/8/1/7/htm.
27. Reid F, Harrigan M. An analysis of anonymity in the bitcoin system. *Security and privacy in social networks 2013*, pp. 197–223. Springer, New York, NY. https://users.encs.concordia.ca/~clark/biblio/bitcoin/Reid%202011.pdf.
28. Beck R. Beyond bitcoin: The rise of blockchain world. *Computer.* 2018; 51 (2):54–58. https://www.computer.org/csdl/magazine/co/2018/02/mco2018020054/13rRUwx1xL2.
29. Androulaki E, Karame GO, Roeschlin M, Scherer T, Capkun S. Evaluating user privacy in bitcoin. *International Conference on Financial Cryptography and Data Security 2013 Apr 1*, pp. 34–51. Springer, Berlin, Heidelberg. https://link.springer.com/chapter/10.1007/978-3-642-39884-1_4.

30. Tama BA, Kweka BJ, Park Y, Rhee KH. A critical review of blockchain and its current applications. *2017 International Conference on Electrical Engineering and Computer Science (ICECOS) 2017 Aug 22*, pp. 109–113. IEEE. https://ieeexplore.ieee.org/document/8167115.
31. Zaghloul E, Li T, Mutka MW, Ren J. Bitcoin and blockchain: Security and privacy. *IEEE Internet of Things Journal*. 2020; 7 (10):10288–10313. https://ieeexplore.ieee.org/document/9122595.
32. Khan FA, Asif M, Ahmad A, Alharbi M, Aljuaid H. Blockchain technology, improvement suggestions, security challenges on smart grid and its application in healthcare for sustainable development. *Sustainable Cities and Society*. 2020;55:102018. https://www.sciencedirect.com/science/article/abs/pii/S2210670720300056.
33. Zhu L, Zheng B, Shen M, Yu S, Gao F, Li H, Shi K, Gai K. Research on the security of blockchain data: A survey. *arXiv preprint* arXiv:1812.02009. 2018. https://arxiv.org/abs/1812.02009.
34. Zhang R, Xue R, Liu L. Security and privacy on blockchain. *ACM Computing Surveys (CSUR)*. 2019; 52 (3):1–34.
35. Lin IC, Liao TC. A survey of blockchain security issues and challenges. *IJ Network Security*. 2017; 19 (5):653–659. http://ijns.jalaxy.com.tw/contents/ijns-v19-n5/ijns-2017-v19-n5-p653-659.pdf.
36. Lin W, Yin X, Wang S, Khosravi MR. A Blockchain-enabled decentralized settlement model for IoT data exchange services. *Wireless Networks*. 2020; 30:1–5. https://link.springer.com/article/10.1007/s11276-020-02345-9.
37. Ma Y, Sun Y, Lei Y, Qin N, Lu J. A survey of blockchain technology on security, privacy, and trust in crowdsourcing services. *World Wide Web*. 2020; 23 (1):393–419. https://link.springer.com/article/10.1007/s11280-019-00735-4.
38. Khan C, Lewis A, Rutland E, Wan C, Rutter K, Thompson C. A distributed-ledger consortium model for collaborative innovation. *Computer*. 2017; 50 (9):29–37. https://www.researchgate.net/publication/319995681_A_Distributed- Ledger_Consortium_Model_for_Collaborative_Innovation.
39. Cheng J, Xie L, Tang X, Xiong N, Liu B. A survey of security threats and defense on Blockchain. *Multimedia Tools and Applications*. 2020; 3:1–30. https://www.researchgate.net/publication/330125746_A_survey_of_blockchain_from_security_perspective.
40. Huang H, Kong W, Zhou S, Zheng Z, Guo S. A survey of state-of-the-art on blockchains: Theories, modelings, and tools. *ACM Computing Surveys (CSUR)*. 2021; 54 (2):1–42. https://dl.acm.org/doi/10.1145/3441692.
41. Dasgupta D, Shrein JM, Gupta KD. A survey of blockchain from security perspective. *Journal of Banking and Financial Technology*. 2019; 3 (1):1–7. https://doi.org/10.1007/s42786-018-00002-6.
42. Honar Pajooh H, Rashid M, Alam F, Demidenko S. Multi-layer blockchain-based security architecture for internet of things. *Sensors*. 2021; 21 (3):772. https:// doi.org/10.3390/s21030772.
43. Nartey C, Tchao ET, Gadze JD, Keelson E, Klogo GS, Kommey B, Diawuo K. On blockchain and IoT integration platforms: current implementation challenges and future perspectives. *Wireless Communications and Mobile Computing*. 2021; 2021:6672482. https://doi.org/10.1155/2021/6672482.
44. Zhang Z, Song X, Liu L, Yin J, Wang Y, Lan D. Recent Advances in blockchain and artificial intelligence integration: Feasibility analysis, research issues, applications, challenges, and future work. *Security and Communication Networks*. 2021; 2021:9991535. https://doi.org/10.1155/2021/9991535.

45. Idrees SM, Nowostawski M, Jameel R, Mourya AK. Security aspects of blockchain technology intended for industrial applications. *Electronics*. 2021; 10 (8):951. https://kdoi.org/10.3390/electronics10080951.
46. Nabben K. Blockchain security as "people security": Applying sociotechnical security to blockchain technology. *Frontiers in Computer Science*. 2021; 2:62. https://doi.org/10.3389/fcomp.2020.599406.
47. Algarni S, Eassa F, Almarhabi K, Almalaise A, Albassam E, Alsubhi K, Yamin M. Blockchain-based secured access control in an IoT system. *Applied Sciences*. 2021; 11 (4):1772. https://doi.org/10.3390/app11041772.
48. Bharati TS. Security and privacy of internet of things. *International Journal of Innovative Technology and Exploring Engineering(IJITEE)*. 2019; 8 (8):2740–2743.
49. Bharati TS. Security enhancement and privacy preserving of big data. *International Journal of Innovative Technology and Exploring Engineering (IJITEE)*. 2019; 8 (10):1754–1758.
50. Bharati TS. Internet of Things (IoT): A critical review. *International Journal of Scientific & Technology Research (IJSTR)*. 2019; 8 (9): 227–232.
51. Bharati TS. Challenges, issues, security and privacy of big data. *International Journal of Scientific & Technology Research (IJSTR)*. 2020; 9 (2):1482–1486.
52. Bharati, T.S, Kumar R. Enhanced key management for mobile adhoc networks. *International Journal of Engineering Science and Computing (IJESC)*. 2016; 6 (4):4184–4187.

# 2 Internet of Things and Industrial Internet of Things Powered by Blockchain

## *A Study on Unleashing Potential*

*S Sandhya, M Senthil Kumar,*
*B Chidhambararajan, and S K Susee*

## INTRODUCTION

The Internet of Things (IoT) is an effective internet and managerial setting that works together to create a world that is intelligently interconnected [1]. IoT users can broadcast the data that they place on the server where it can be used intelligently. At present, IoT is being used in many contexts, such as corporations, industries, environmental engineering, mobile devices, and governance. IoT also helps the transportation, healthcare, agriculture, energy, and environmental management industries. Additionally, IoT supports many commercial initiatives in the fields of services, manufacturing, logistics, and food production. Academics and businesses have paid a lot of attention to Industrial IoT (IIoT), which could be very important in the upcoming conversion of industrial schemes. IoT plays a vital role in Industry 4.0, and it is also one of the most important topics in academics and politics. The major difference between IIoT and IoT is that whereas IIoT is frequently applied in applications like smart factories and smart manufacturing, IoT environments frequently feature a variety of automation equipment and industrial devices.

IIoT provides intelligence and connectivity to industrial systems through the help of drives, sensing devices, a pervasive network, and processing power. Implementing IIoT is intended to increase manufacturing efficiency and business productivity, lessen device downtime, and raise the caliber of the final output. Specifically, IIoT features the following characteristics: decentralization of IIoT systems, a variety of IIoT-based applications, schemes, heterogeneity of IIoT-based information, and network strain [2].

DOI: 10.1201/9781003428886-3

Advancements in blockchain technology increase the chances of solving IIoT difficulties. Bitcoin's core technology is blockchain. Blockchain makes it possible for computer programs to send and receive messages or keep transactions in a widely accepted (peer-to-peer) fashion. Blockchain is swiftly gaining popularity and is mostly used for distributed storage, smart contract applications, and digital assets [3]. Applications that use the blockchain for IIoT include storage attributes (such as temperature, humidity, or location deviations) and creating block challenge ledgers that are understandable only by certain users, like particular contributors of the supply chain. Blockchain technology has transmuted from an unworkable approach to a reliable approach to problem-solving.

Smart contracts (SCs) have been used in digital economics based on blockchain 2.0 in an automatic way, whereas blockchain 1.0 is generally associated with cryptocurrencies and overhead (for example, Bitcoin). The focus of more recent blockchain 3.0 is on meeting the demands of the digital world, such as Industry 4.0's concerns with smart cities. IoT continuously energizes Industry 4.0. Industry 4.0 is associated with IIoT. Though one may regard IIoT as an improvement of IoT, it also involves a further focus on IoT use in industry. Additionally, Industry 4.0 is frequently associated with ethical manufacturing workplace groups.

The blockchain's internal secular and spatial data can be found and verified via traceability. Additionally, each data block stored within a blockchain has a suitable timestamp added to it, ensuring data discovery.

With IIoT equipment and physical equipment, interoperability may exchange data. Building an overlay P2P network on top with persistent access across many IIoT methods will enable interoperability within the blockchain composite layer.

The potential for IIoT to cooperate without the interference of a trusted third party (TTP) is discussed in autonomous interactions. Blockchains enable smart contracts to attain this autonomy. Particularly, contract clauses encircled by smart contracts will be habitually enforced after a specific circumstance is met.

## Motive

Cloud-based manufacturing (CBM) is a new phenomenon in the most recommended pattern for production. An isolated group of flexible resources for production can be quickly given with a minimum of service provider administrative effort or contact thanks to CBM's constant, suitable, whenever-needed network connectivity. A decentralized, peer-to-peer IIoT system called BIIoT is suggested. A central CBM operator, the BIIoT system improves the usefulness of older CBM systems, especially when integrating outdated floor equipment in cloud environments. Additionally, resources that are meant for production are enabled by CBM on an on-demand basis, necessitating the usage of trustworthy distributors in transactions between users unable to contract for production services.

Though BIIoT stands as an approval to IIoT, a number of problems need to be overcome before it can be successfully implemented and its potential fully realized. As a result, the purpose of this chapter is to offer a thorough overview of the architecture that reports difficulties and explores BIIoT-related research issues.

### Contributions and Comparison with the Survey

The pairing of blockchain with IIoT is discussed in a number of published articles. For instance, Wen et al. describe a blockchain-based information distribution system for supply chain (SC) networks using IIoT. The system connects IIoT nodes to the blockchain, monitors IIoT nodes, and uses smart contracts to store concurrent data within the network. In addition, a blockchain-based supply chain design is suggested. They suggested a design that proposes supply chain collaboration options that link various commodities. An information system for fine-grained distribution is proposed inside this supply chain concept.

Li et al. suggested a consortium blockchain-based energy blockchain secure power business model [5]. This kind of blockchain is often used in international peer-to-peer (P2P) power purchase agreements since it removes the middleman. Additionally, it decreases the effects of contract restriction caused by delays during the verification of contracts on the energy blockchain. Many researchers have proposed a credit-based pricing structure to encourage continuous and short-term power purchase agreements. The list of blockchain platforms used in IIoT is provided by Teslya et al. [6]. The most common consensus techniques are examined, together with the unique traits of public blockchain and private blockchain.

Through the analysis of prior work, one aims to:

1. Present a conceptual preamble on IoT, IIoT, and blockchain.
2. Permit the examination of the viability of integrating IIoT into blockchains.
3. Create a facility of consultations in a perceptive way on how to address technological challenges while implementing BIIoT.

   Here is a quick summary of the contributions:
4. An outline of the key elements of IoT was given in the introduction. Additionally, a summary of IoT difficulties is provided.
5. In addition, a concise overview of IIoT is given, along with a list of its main properties.

In addition, the IIoT research problems are listed.

## IOT: AN OVERVIEW

The term "Internet of Things (IoT)" refers to a collection of "things" that are comprised of software, electronics, actuators, and sensors and connected via the web in order to exchange information. IoT nodes are made up of processing power and sensor devices, which are common in numerous industries. Figure 2.1 shows the devices that are connected through IoT. IoT home automation refers to the capacity to automate household appliances using digitally controlled, network-connected systems to uphold the safety of the populace. An IoT system can monitor and control anything remotely, including lights, through the component known as a smart grid [7].

As a result of severe weather and earthquakes, electricity influxes like traffic signs, traffic congestion, parking, and predictions of things are affected. IoT devices

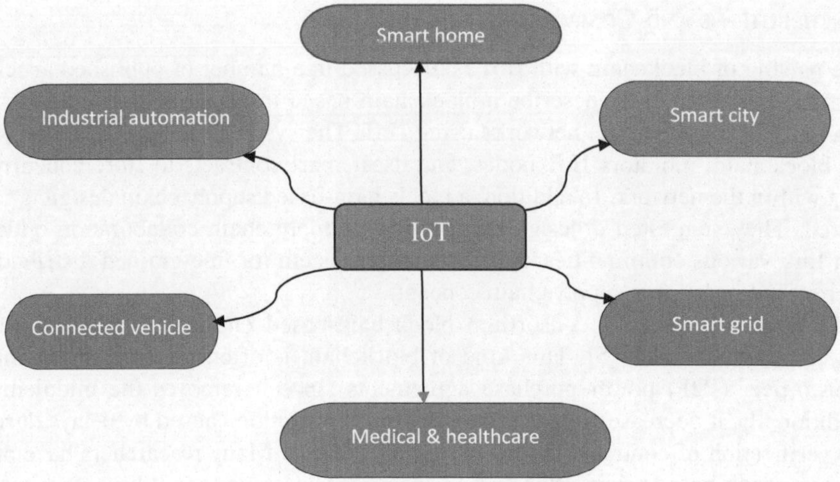


**FIGURE 2.1** IoT connectivity.

with sensors are used to track the whereabouts in real time of medical equipment like wheelchairs, oxygen concentrators, defibrillators, and other security tools. Forecasting is one of automotive IoT's most promising applications. The technology gathers performance information from the connected car's chips and sensors, which can then be analyzed in the cloud and predicted before repair is needed [4].

According to 2013 research from the International Data Corporation (IDC), the market for IoT equipment is expected to grow quickly to 41 billion units by 2020 and reach a value of $8.9 trillion. The absence of human responsibility distinguishes the Internet of Things from traditional internet usage. IoT nodes can generate data on a person's activity, analyze it, and choose the best course of action. IoT services have made it possible to increase human quality of life. However, during the extraction of human–interface data via IoT devices and sensors, confidentiality and data protection are of the utmost importance.

## IOT OBSTACLES

Battery life and streamlined computational needs are the two criteria that IoT devices must meet above all else.

## DURATION OF A BATTERY

Many IoT nodes are positioned in remote areas without access to power. The nodes only have a limited amount of power to carry out the intended function and necessary safety commands, which often deplete the node's battery power. There are three practical strategies to resolve the battery life issue [8]. The most crucial strategy is to use the minimal amount of security requirements on the node, which is not advised, especially during the handling of sensitive information. The second strategy involves increasing the battery's capacity. The majority of IoT nodes are planned and constructed to be lightweight and with less volume, which leaves less room for a larger battery. The last strategy utilizes natural sources (such as heat, light, wind, and vibration) to generate a sufficient amount of power.

## THE DIFFICULTY OF CONFIDENTIALITY AND SAFETY

Although the safety solutions now being utilized on the web are deemed to be gainful and viable, applying them to IoT will not provide the desired results for the following purposes:

- Nodes typically lack the computing power needed to carry out complex security measures.
- Nodes used in many IoT systems might not be physically protected, leaving them open to malicious users.
- It may not always be possible to join via a node. For instance, a node might be in "sleep mode" to conserve power, or it might be mobile and not accessible at the moment.

## THE SUSTAINABILITY CHALLENGE

Conventional processing nodes can be improved or modified by customers, but it is not suitable when it comes to IoT devices like cars, ovens, or fridges. Update cycles of IoT, however, raise concerns regarding the durability of such nodes in a linked society. For example, will the nodes upgrade their operating system over time, or will this result in a fragmented network with insecure hardware?

When installing IoT instances, a node can also be a part of advanced communication systems or a remote installation node that takes the place of labor-intensive, expensive, or impractical techniques. While building a bridge over a river, for instance, temperature and earthquake sensors with fire alarms as well as pollution finding sensors are commonly installed.

## THE DIFFICULTY OF THE TRUST APPROACH

The improved communication opportunities between the virtual and real worlds are maybe the most important improvement brought forth by IoT. IoT has been conceptualized to include common machines as nodes, which will greatly alter the course of our lives. A reliable technique is sought to enable successful and fruitful communication between these nodes with little or no interference from individuals. The system must take into account valid sensor and actuator performances, which occasionally occur during severe crashes. Communication should be permitted as well with the help of reward–responsibility-based techniques [9].

## IIOT DIFFICULTIES

IIoT ensures the connectivity of diverse things by using a variety of actuators. Sensors and software systems are installed to detect and gather data from the actual environment, and those devices should generate the actions in a subsequent manner. The unique features of IIoT provide a chance to study a number of examination issues. Information and communication technology (ICT) advancements in recent years have made it possible to overcome some of the fundamental limitations of IIoT. For example, interactions with ambient backscatter can help IIoT devices and increase power. Additionally, by shifting the process-intensive tasks to edge servers, mobile edge computing can increase the functionality of the IIoT device. Additionally, current advancements in blockchain technologies bring difficulties, including inadequate interoperability, safety, and confidentiality vulnerabilities.

## BLOCKCHAIN TECHNOLOGY: AN OVERVIEW

### BLOCKCHAIN

A distributed ledger that spreads throughout an entire distributed system is referred to as a blockchain. With decentralized permission, blockchains can enable contracts to take place and validate collectively distributed things. Blockchains can, therefore, perform decentralized contract confirmation, leading to lower costs and traffic operation in the centralized organization. Additionally, as every node in the network keeps a record of the completed transactions inside the blockchain, all transactions saved in blockchains are impenetrable. Meanwhile, cryptographic techniques ensure the data block's integrity inside the blockchain. As a result, blockchains can ensure that communications are not lost [10]. All contracts can also be tracked thanks to the significant timestamp that was assigned. A blockchain is made up of neatly organized parts called blocks that have headers with transaction information. By using a predecessor's hash, the predecessor reflects the header and metadata-related information. Within the initial block, known as the genesis block, which has no old model or method, the first position is encoded in a hard sense. Merkle trees are used for hashing the transactions that are based on Ethereum and Bitcoin. A blockchain instance with blocks and repeated chains is depicted. With the help of the parent block hash value, every block in a blockchain points to its immediately previous block (referred to as the parent block). For instance, block i contains the block i – 1 hash.

1. Block version: The validation rules to be followed.
2. Hash of parent block: represents the 256-bit hash value of the first block.
3. Merkle tree root hash: Holds the hash values for each transaction contained within the block.
4. Nonce: A 4-byte value that typically starts with 0 and rises for all hash calculations.
5. Timestamp: Starts on January 1, 1970, and has time in seconds in UTC.
6. A true block hash point threshold is nBits.

### Consensus Algorithms

The one major advantage of blockchain techniques is the ability to verify the constancy of the block in decentralized, despondent settings that need the authorization of a trusted third party. However, a consensus is required in distributed contexts. The agreement may also be biased in favor of good devices, therefore, it must be obtained on a newly constructed block. Consensus methods are often used to attain this level of reliability in decentralized systems. Practical Byzantine fault tolerance (PBFT), proof of stake (PoS), and proof of work (PoW) are common consensus algorithms. Consensus algorithms are an essential piece of the puzzle for effective blockchain operating rules since they safely update duplicated distributed conditions.

### Workflow of Blockchains

The following shows a functional instance of the blockchain. The method of money transfer is as follows:

Step 1: Transfer cash from person A to person B.
Step 2: A Bitcoin wallet is used to begin the contract at a system. The transaction comprises the recipient's address, the sender's wallet, and the total amount of currency.
Step 3: A block represents the online transaction. The block is then advertised to every node in the network using the next three steps.
Step 4: Network members confirm the transaction's validity.
Step 5: After that, a verified transaction is added to the top of the contact's chain, creating a new block inside the blockchain.
Step 6: The money is transferred from person A to person B.

## SMART CONTRACT

Blockchain-based smart contracts are programmable programs that control transactions under certain conditions. Smart contracts are created based on blockchain technology, which raises great interest among the stakeholders and several enterprises. Due to this interest, it has become a technology that is welcomed in enabling the automation of digital processes among the parties to be verifiable and enforceable.

Smart contracts have their own blockchain accounts and addresses, which is why they are sometimes referred to as "independent agents".

In order to complete the predetermined criteria, the contract may gain ownership of tokenized assets while designating the contributor as the employee. IoT and financial services are just two of the many areas that the smart contract has the potential to change. Smart contracts are frequently used on a blockchain network to carry out and advance aims, such as:

1. Allowing transactions to be initiated automatically by specific actions. This function can take many different forms, such as automatic transmissions at regular intervals or transmissions in response to earlier transmissions.

2. They permit so-called multisignature transactions, in which a transaction is only released when specified or most participants confirm that they have signed it.
3. Allowing storage of specific application data, such as membership records, Boolean values, or list statuses.
4. They make other people's smart contracts useful.

For instance, inheritances are frequently included in Ethereum's smart contracts, where a single contract can refer to the goals specified therein.

## BLOCKCHAIN SYSTEM TAXONOMY

The three categories of blockchains are public, private, and consortium . With permissionless public blockchains, everyone can participate in the distribution of new blocks and use of blockchain data. Public blockchains are described as decentralized since everyone can keep a copy of the blockchain by working together to confirm new blocks. All nodes in the network are a part of permissioned blockchains, known as private blockchains. Private blockchains are approved by a single company and accessed by a coordinated person or team, and the distributed database holds data exchanges between different people or departments. Blockchains used by consortiums resemble private blockchains. It is a network with permissions. Consortium networks connect many different organizations and allow the participants to communicate clearly.

## INTEGRATING BLOCKCHAIN WITH IIOT

IIoT systems must overcome a number of obstacles, including poor interoperability, heterogeneity, resource limitations, confidentiality concerns, and security flaws. Blockchain technologies can balance IIoT systems by enhancing confidentiality and safety while also enhancing interoperability. Blockchains can also make the IIoT system more dependable and quantifiable. BIIoT is the name given to the result of combining blockchain and IIoT.

## BIIOT ARCHITECTURE

The blockchain composite layer is used in this design as a distributor between IoT and industrial applications. This design offers the following two important advantages: user access to blockchain-based services and an idea from the IIoT's fundamental layers. The blockchain composite layer is used to provide heterogeneity to the lower layers and sufficient API in order to support industrial demands. Therefore, the concept made possible by the blockchain composite layer can help reduce the difficulty of creating industrial applications. The five sublayers of the blockchain composite layer's descriptions are also displayed. In a typical IoT architecture and framework, the subscheme layer concept from bottom to top is used.

The perception layer allows various IoT devices to sense and collect data from their immediate ambiances. These devices include actuators, sensors, RFID tags, controllers, and smart meters.

The communication layer helps to build an industrial network. Many IoT devices can connect by using the WAP–Wi-Fi Access Point, Base Station (BS), Macro BS, and gateways for IoT. Several communication protocols, including Near Field Communications (NFC), Bluetooth, Wireless Highway Addressable Remote Transducer (WirelessHART), and Low Power Wide Area Networks (LPWAN) techniques with LoRa, Sigfox, industrial Ethernet, and Narrowband IoT (NB-IoT), are capable of connecting IoT sensors.

A variety of industrial applications can be broadly maintained by using IoT. The food sector, manufacturing, smart grid, and supply chain are among the common industrial applications.

## Deployment of BIIoT

It is difficult to save the whole blockchain at IIoT nodes due to imitation IIoT nodes. The real BIIoT distribution is of maximum importance. There are two types of storage: partial storage and complete storage. With partial storage, just a portion of the information blocks are locally kept. Lightweight nodes is another name for partial storage nodes. Complete storage saves the entire blockchain. Complete nodes refer to entire storage nodes (such as edge servers and cloud servers). Since storing many complete blockchains requires a lot of storage space and current processing power to overcome consensus problems, a complete node has enough processing power. Alternatively, IoT devices with limited resources are small devices that guarantee transaction faithfulness without extracting. IIoT devices perform the swap of partial blockchain information through D2D connections, and it is referred to as direct communication with the IIoT devices. Also, the IIoT devices can transfer the blockchain-based information that is stored in the edge servers with the help of MBS and IIoT-based gateways, and this type of communication is referred to as direct communication among blockchain and IIoT. Hybrid communication between edge servers and blockchain [11]

## Applications for BIIoT

The fusion of IIoT and blockchain is a new, rapidly expanding trend. This chapter provides an overview of BIIoT applications because blockchain technology can help with overcoming IIoT problems. Blockchains have a wider range of applications than other technologies. This study is mostly about industrial applications.

## Food Sector

BIIoT makes a commodity's life cycle more visible, especially in the food sector. Food product traceability (FPT) is essential for ensuring that the food is protected. The entire food supply chain is guaranteed by FPT along with IoT. For instance, a company that manufactures food might be recommended by several suppliers. The entire food manufacturing business must therefore digitize information from sources. Blockchain technology can ensure the food's provenance and tracking capabilities through this process. An RFID and blockchain-based supply chain platform in China connects farmers and food producers. The purpose of this system is to make sure that the food supply chain's data can be tracked.

Further, blockchain technology can help increase food protection by making trackable food production more accessible. It demonstrates how integrating blockchain into the food supply chain can enable clients to trace the entire process of food production. Researchers have offered a blockchain application case for the industry of natural coffee in Colombia. It provides a facility to track the food protection system that is supported by blockchain technology and electronic product code (EPC) IoT tags. Through smart contracts, this technology can specifically stop the interception and disclosure of confidential information. To demonstrate its effectiveness, a model of the proposed architecture has been put into action.

## Smart Manufacturing

The manufacturing sector is moving from mechanical to smart techniques. Big data analytics is crucial to this advancement in the manufacturing industry. A huge amount of information is produced during each stage of a commodity's life cycle, including the procurement of raw materials, designing, sharing, manufacturing, after-sales repair, and selling of manufactured goods. However, information aggregation and analytics are required due to the fragmented nature of the production data. The interoperability problem is solved by BIIoT by integrating the IIoT-based systems through P2P networks, which facilitates the sharing of information among the various industrial sectors. For example, several distributed blockchains, with each supporting a distinct industry, are commonly constructed to support diverse industries. Through BIIoT, security in smart manufacturing can be enhanced [12].

The majority of enhanced firmware is physically installed at IIoT nodes after being physically imported from a central server. The firmware updates must be organized in a distributed IIoT, which is expensive and time-consuming. The work offers a standard firmware upgrade option that uses blockchain technology and smart contracts. The entire industrial network is organized around smart contracts that handle firmware improvement strategies. Smart contracts are implemented by design and can be used by nodes to install and import firmware hashes. A decentralized blockchain-based autonomous manufacturing solution was put up to offer stronger security than the typical centralized architecture. In addition, blockchain was used to develop a mobile-based crowd-sensing system to address the issue of information in smart manufacturing.

## Healthcare Sector

The rapid development in the field of digital technology plays an important role in the progress of healthcare services, which has the potential to free clinical data and offer a more practical means of enhancing healthcare outcomes and processes. Because hospitals have limited resources, healthcare services currently face new obstacles. The probability of developing remote healthcare services is increased by recent developments in wearable medical technology and big data analytics in healthcare information. As a result, hospital resources are regularly and effectively released. For instance, elderly people who live in their homes wear medical devices on their bodies. These gadgets continuously measure and collect health data, including blood pressure, heart rate, and sugar level.

Through healthcare networks, medical teams and lab assistants can access medical information at any time and from any location. However, using medical data raises concerns about security, privacy, and confidentiality. Healthcare information confidentiality and safety are challenged by network heterogeneity and vulnerabilities prevalent in medical technology. Blockchain-based healthcare network integration can successfully address issues with healthcare data security and confidentiality. For instance, blockchain technology might be used to protect healthcare data stored on cloud servers. Notably, to support real-time patient observation, the healthcare data generated by medical sensors is on a routine basis, and based on the smart contract, it is broadcast to the system. Using blockchains, confidentiality is typically maintained throughout the entire process.

Digital signatures can identify the owner of healthcare information and verify the validity of the information. This system can also safeguard the owner's right to privacy when it comes to medical information. In order to offer confidentiality and secrecy, the present effort combines blockchain-based mobile edge computing with IoT to establish a framework for controlling in-home therapy. It also provides traceability and possibilities for patients who have contracted contagious viruses such as COVID-19, Middle East respiratory syndrome, and severe acute respiratory syndrome. Notably, patients using IIoT technology can be consecutively tracked in a manner akin to quarantine while maintaining patient confidentiality.

## Automotive Industry

Industry 4.0 refers to the post-digitalization stage that suffers from cyberattacks and several issues related to safety and operability. Due to advancements like IIoT integration along with IoT connected to electric, hybrid, and self-driving smart cars, the automotive industry is one of the most technologically advanced sectors. These issues have just been approved for dissemination to the many different parties who participated in the vehicle's lifespan. The utilization of enormous sensors and information technology is made possible by Industry 4.0, which benefits from developments in several fields, including the expansion of cutting-edge technologies and the rise in connectivity and computational power.

Multiple services and features are made possible by the expanding capacities offered by complex diverse linked and autonomous networked organisms. They do, however, exist due to the danger of malicious assaults or other dangers, which call for more effective cybersecurity. Furthermore, open protection is promised when the restricted organisms are based on vehicle-associated systems, making strong cybersecurity an absolute necessity. The automotive sector may make use of BIIoT technologies to alleviate this problem.

## Fuel Sectors: Oil and Gas

In the upcoming 20 to 30 years, the global energy market will be dominated by fuels like oil and gas. Due to the importance of oil and gas resources to the energy sector, cutting-edge drilling techniques, digital platforms for the marine industry, and intelligent oil and gas fields have all benefited from the industry's recent quick embrace of technology. The oil and gas business frequently integrates automation, digitalization, and intellectualization. It stands out due to its limited effectiveness, high cost, significant risk, prolonged duration, and outdated organizational structure.

Based on the segmentation of the market, the fuel industry is classified into three types: downstream, midstream, and upstream. While downstream and midstream alludes to sales and storage of energy sources, upstream describes the creation and examination of oil and gas.

Regarding the mentioned problems, the oil and gas sector needs to adopt a new style of management. The oil and gas business has big potential for using blockchain. The debut of Bitcoin in 2008 created a great impact and influence in the advancement of the blockchain technology. The oil and gas sector has adopted a wait-and-see attitude in the interim.

## Trade Supply Chain Industry (TSCI)

The final step in the trade supply chain is the distribution of the goods to the client. The trading technique includes a valid combined contract between the buyer and the seller that outlines all of the conditions necessary for a successful trade. These scenarios, for instance, could determine the best method of product transport. The trading process is usually examined by a third party. The main responsibility of this organization is to make sure that each party complies with the terms of the contract. If a dispute arises between a buyer and a seller, it will be resolved by the regulatory body. Customers who use the Amazon trade platform, for instance, can alert the platform's management in case they get goods that do not satisfy the merchants' advertised descriptions.

## Open Research Issues in BIIoT

Blockchain and IIoT together offer a variety of chances to advance the sector. Before the capability of BIIoT is regularly and fully employed, numerous problems need to be overcome. Numerous key obstacles to IIoT integration into the blockchain are examined in the following sections.

## Privacy Leakage

There are many protection layers built into blockchain technologies to prevent unauthorized access to transaction-based reports maintained in blockchains. Bitcoin transactions are made using IP addresses rather than actual customer identities, which ensures personal confidentiality. Even though these protection measures are inconsistent and insufficient, one-time accounts (OTAs) can be created under Bitcoin to help users maintain their anonymity. For instance, it has been found that consumer pseudonyms are regularly broken by inferring from and learning about the various transactions connected to a single general user. In addition, the entire blockchain's transaction data storage space can still have an impact on the possibility of confidentiality leakage, as noted.

## Security Vulnerability

IIoT can increase security through encryption and signature creation by blockchains. BIIoT continues to work actively on protection due to the weaknesses of both blockchain and IIoT systems. Building wireless networks in industrial contexts is becoming more popular since wireless communication organizations can be measured and are practical. Security breaches like jamming, submissive overhearing, and recurrent attacks are caused due to the open wireless channel. Additionally, due to the resource restraints of IIoT nodes, conventional extreme weighted encryption methods might not be applicable in the IIoT. In dispersed locations, it can be difficult to keep an eye on the keys as well.

## Resource Constraints

Numerous IIoT nodes with constrained resources are present. RFID tags, sensors, and smart meters, for instance, have lessened network link ability, constrained storage, and decreased computational power. Additionally, the decentralized consensus techniques employed by blockchains typically consume more energy and need more computational power. For instance, it has been proven that Bitcoin's PoW uses a significant amount of electricity. Therefore, the consensus technique, which necessitates a large power expenditure, may not be practical for low-energy IIoT nodes. On the other hand, the sheer volume of blockchain data makes it impractical to implement blockchains completely across the IIoT. For instance, the size of blockchain-based Bitcoin had crossed 185 GB by the end of September 2018. An IIoT node cannot fully save the entire blockchain.

## Scalability

Scalability limits the usage of the blockchain in the context of extensive IIoT. It is common practice to measure the quantity of concurrent workloads, the size of the IIoT nodes, and the throughput of transactions per second. Failure transaction issues have disappointed many blockchain initiatives. For instance, Bitcoin's evolution to seven transactions per second is demonstrated. PayPal can process 170 transactions per second, and VISA can produce close to 2000 transactions per second. Due to weak measurability, the Bitcoin blockchain may not be appropriate for IIoT. As a result, applications like IIoT that require a high rate of transactions may not be suitable for current blockchain platforms.

## CONCLUSION

The Industrial Internet of Things is now facing many difficulties, such as inadequate interoperability, heterogeneity, security and confidentiality flaws, and resource constraints. IIoT challenges can be solved by utilizing current blockchain technologies, which improve privacy, interoperability, trackability, safety, and dependability. In order to undertake the investigation described in this chapter, blockchain and IIoT were combined. The term "BIIoT" refers to such a marriage of blockchain and IIoT, and this study conducts a thorough survey of the literature published in this area. The main topics that have been studied and presented include the Industrial Internet of Things, blockchain technology, and the Internet of Things. The chapter also assesses the applications of BIIoT and illustrates BIIoT architecture. Detailed illustrations of BIIoT-capable applications have been provided, along with a list of BIIoT open research issues. Due to their high centralization and single point of failure, IIoT security systems, as they currently exist, are inefficient. However, IIoT that is based on blockchain is particularly resistant to external hacking and other threats. The same data is kept on each blockchain node, so there is no chance of data loss. Future work will put the BIIoT system into practice and demonstrate it for practical solutions like on-demand manufacturing and device self-service.

## REFERENCES

1. A. AlAbdullatif, K. AlAjaji, N. S. A. Serhani, R. Zagrouba, M. AlDossary, "Improving an Identity Authentication Management Protocol in IIoT", *2nd International Conference on Computer Applications & Information Security (ICCAIS)*, Riyadh, Saudi Arabia, July 2019.
2. I. Makhdoom, M. Abolhasan, J. Lipman, R. P. Liu, W. Ni, "Anatomy of threats to the Internet of Things", *IEEE Communications Surveys & Tutorials*, vol. 21, no. 2, pp. 1636–1675, 2018. DOI: 10.1109/ COMST.2018.2874978.
3. F. Tariq, M. Anwar, A. R. Janjua, M. H. Khan, A. U. Khan, N. Javaid, "Blockchain in WSNs, VANets, IoTs and Healthcare: A Survey", *Artificial Intelligence and Network Applications, WAINA 2020. Advances in Intelligent Systems and Computing*, vol. 1150, Springer, Cham, 2020.
4. H. G. Do, W. K. Ng, "Blockchain-based System for Secure Data Storage with Private Keyword Search", *2017 IEEE 13th World Congress on Services*, Honolulu, HI, 2017. DOI: 10.1109/ SERVICES.2017.23.
5. Z. Li, J. Kang, R. Yu, D. Ye, Q. Deng, Y. Zhang, "Consortium Blockchain for Secure Energy Trading in Industrial Internet of Things", *IEEE Transactions on Industrial Informatics*, vol. 14, no. 8, pp. 3690–3700, August 2018. DOI: 10.1109/TII.2017.2786307.

6. N. Teslya, I. Ryabchikov, "Blockchain Platforms Overview for Industrial IoT Purposes", *2018 22nd Conference of Open Innovations Association (FRUCT)*, Jyvaskyla, Finland, pp. 250–256, 2018. DOI: 10.23919/FRUCT.2018.8468276.
7. Q. Lu, X. Xu, "Adaptable Blockchain-based Systems: A Case Study for Product Traceability", *IEEE Software*, vol. 34, no. 6, pp. 21–27, November/December 2017.
8. P. K. Sharma, S. Singh, Y.S. Jeong, and J. H. Park, "Distblocknet: A Distributed Blockchain-Based Secure SDN Architecture for IoT Networks", *IEEE Communications Magazine*, vol. 55, no. 9, pp. 78–85, September 2017. DOI: 10.1109/ MCOM.2017.1700041.
9. A. Dorri, S. S. Kanhere, R. Jurdak, "Towards an Optimized BlockChain for IoT", *IEEE/ACM Second International Conference on Internet-of-Things Design and Implementation (IoTDI)*, Pittsburgh, PA, June 2017.
10. M. Conti, S. K. E, C. Lal, S. Ruj, "A Survey on Security and Privacy Issues of Bitcoin", *IEEE Communications Surveys & Tutorials*, vol. 20, no. 4,pp. 3416–3452, 2018. DOI: 10.1109/COMST.2018.2842460.
11. G. Sagirlar, B. Carminati, E. Ferrari, J. D. Sheehan, E. Ragnoli, "Hybrid-IoT: Hybrid Blockchain Architecture for Internet of Things – PoW Sub-Blockchains", *IEEE Conference on Internet of Things, Green Computing and Communications, Cyber, Physical and Social Computing, Smart Data, Blockchain, Computer and Information Technology*, Halifax, NS, Congress on Cybermatics, 2018.
12. J. Leng, D. Yan, Q. Liu, K. Xu, J. L. Zhao, R. Shi, L. Wei, D. Zhang, X. Chen, "ManuChain: Combining Permissioned Blockchain With a Holistic Optimization Model as Bi-Level Intelligence for Smart Manufacturing", *Systems Man and Cybernetics: Systems IEEE Transactions*, vol. 50, no. 1, pp. 182–192, 2020.

# 3 Utilizing Blockchain and Interplanetary File System for Enhanced User Privacy in Secure Data Sharing

*K Balamurugan, M Azhagiri, P A Harsha Vardhini, Edwin Jijo, and Latha Parthiban*

## INTRODUCTION

In the current digital era, the proliferation of data sharing across various platforms and applications has raised significant concerns about user privacy and data security. Sensitive data integrity and confidentiality are jeopardized by hacking, data breaches, and unauthorized access of centralized traditional data storage systems. In response to these challenges, innovative technologies such as blockchain and the InterPlanetary File System (IPFS) have emerged as promising solutions for enhancing user privacy and security in data-sharing environments [1].

Blockchain technology was initially developed as the base infrastructure for virtual currencies like Bitcoin but has now expanded its range to serve a variety of purposes, including safe data sharing. In essence, blockchain can transparently and impenetrably record transactions which is immutable. Blockchain eliminates the need for intermediates and guarantees data integrity by spreading data among a network of nodes and using cryptographic techniques to encrypt transactions. This lowers the possibility of unauthorized access and manipulation [2].

Complementing blockchain technology, IPFS offers a decentralized approach to the sharing of files and their storage. As opposed to conventional file systems, which depend on centralized servers, IPFS utilizes a peer-to-peer network to distribute and retrieve content in a distributed manner. Files stored on IPFS are identified by their unique content hashes, allowing users to access and share data directly from the network without relying on centralized servers. This decentralized architecture

 DOI: 10.1201/9781003428886-4

enhances data availability, resilience, and privacy because there isn't a single control or point of failure [3].

By bringing together blockchain with IPFS, organizations can create an impeccable and privacy-enhancing framework for data sharing. The combination of blockchain's tamper-proof ledger and IPFS's decentralized file storage provides a robust infrastructure for storing, sharing, and accessing sensitive data while preserving user privacy. Blockchain technology ensures the integrity of data transactions, whereas IPFS facilitates secure and efficient content distribution across the network [4].

Automating access control procedures on the blockchain can further improve user privacy. Smart contracts are contracts embedded into the blockchain with preset rules and conditions. Organizations can edit sensitive information for data sharing [5].

The probability of cyberattacks is lowered by the decentralized structure of IPFS and blockchain. Because blockchain and IPFS disperse data over a network of nodes rather than centralized servers, which are vulnerable to hacking and data breaches, it is far more difficult for malicious activity to take place. Additionally, using techniques related to cryptography as encryption and digital signatures enhances the security of data transactions, further mitigating the risk of unauthorized access [6].

In addition to enhancing user privacy and security, the combination of blockchain and IPFS offers other benefits for data-sharing environments. For example, the transparent and immutable ledger of blockchain provides a verifiable record of data transactions, enhancing accountability and auditability. Meanwhile, IPFS's distributed file storage improves data availability and resilience, ensuring that files can be accessed even in the event of network disruptions or failures [7].

The Internet of Things (IoT) is quickly changing the industry and enabling previously unheard-of levels of automation, data analysis, and collecting by linking gadgets, sensors, and machines to the internet. But as IoT networks grow in stature and complexity, scalability, security, and privacy issues confront traditional centralized designs. Blockchain technology has come in as a potential solution for these issues, improving the security, privacy, and effectiveness of IoT networks [8].

Originally intended to serve as the foundational technology that facilitates safe transactions among a network of dispersed nodes, blockchain reduces the danger by ensuring data integrity and doing away with the need for intermediaries by recording transactions in an immutable and tamper-proof manner. Blockchain technology, when combined with IoT devices, can offer a decentralized and safe architecture for controlling access control, data transfers, and device IDs in IoT networks [9].

Updated and upgraded security is an important advantage of using blockchain with IoT. As opposed to systems that store data centrally, which often collect and monetize user data for targeted advertising or other purposes, blockchain and IPFS enable users to retain data control. By data storage on a decentralized network and employing encryption techniques, blockchain, and IPFS make sure that private and inaccessible information stays out of the reach of unauthorized people. Conventional IoT designs are vulnerable to cyberattacks and single points of failure since they frequently rely on centralized servers for data processing and storing. On the flip side, blockchain technology uses decentralized data that is dispersed over a network of nodes, which increases the difficulty of systemic compromise by malicious actors. Additionally, by guaranteeing that only authorized access and editing are available on data stored on the blockchain, blockchain's cryptographic algorithms and consensus methods improve data security [10].

Blockchain offers an open and auditable record of data exchanges, and it strengthens data integrity and trust in IoT networks. Each transaction recorded on the blockchain is timestamped and cryptographically secured, making it tamper-proof and verifiable. This transparency and immutability of data transactions improve accountability and enable stakeholders to trace the origin and history of data in IoT ecosystems, fostering trust and reliability [11].

The other key benefit of integrating blockchain with IoT is improved privacy and data ownership. Traditional IoT architectures often raise concerns about data privacy, as centralized servers have access to huge amounts of sensitive information collected from connected devices. By contrast, blockchain enables users to retain control over their data by storing it on a decentralized and permissioned network. This gives assurance that sensitive data are kept private and inaccessible to unauthorized parties, enhancing user privacy [12].

Blockchain enhances the efficiency and reliability of IoT networks by enabling automated execution of smart contracts with preset terms and conditions. By deploying smart contracts in IoT ecosystems, organizations can automate various processes, such as device provisioning, data sharing, and payments, thereby reducing manual intervention, minimizing errors, and improving operational efficiency [13].

Blockchain-based systems have a significant hurdle in transaction and contract execution performance prohibiting them from competing on a wide scale with present applications. Evaluation of the performance of blockchain technology and the metrics linked with it have been the subject of a significant amount of research. A system's performance is assessed using a procedure known as performance evaluation. Many performance evaluations are based on determining how a change in a dependent variable affects performance.

The layer-based IoT operation is currently classified into three different taxonomies: three-, four-, or five-layered architectures. A closer look at all the taxonomies shows that three layers are the most frequently adopted in most IoT applications. Therefore, the proposed system considers choosing a three-layered architecture [14].

The three-layer architecture is shown in Figure 3.1. The application layer, which defines all IoT applications, is frequently impacted by several types of infiltration, such as malicious codes and cross-scripting assaults. It is contingent upon the application's sensitivity, though. All of the data received from the IoT device is forwarded using the network layer, which is also utilized for transmission. This layer hosts the majority of the communication protocol. There are several ways for interruptions to occur in these layers, including exploitation and storage. The perception layer is responsible for extracting the raw sensory data and transmitting it to the network layer. This layer is typically vulnerable to node capture and eavesdropping replay attacks. Since the goal of the proposed study is to create a secure communication scheme, the network and application layers take care of this [15].

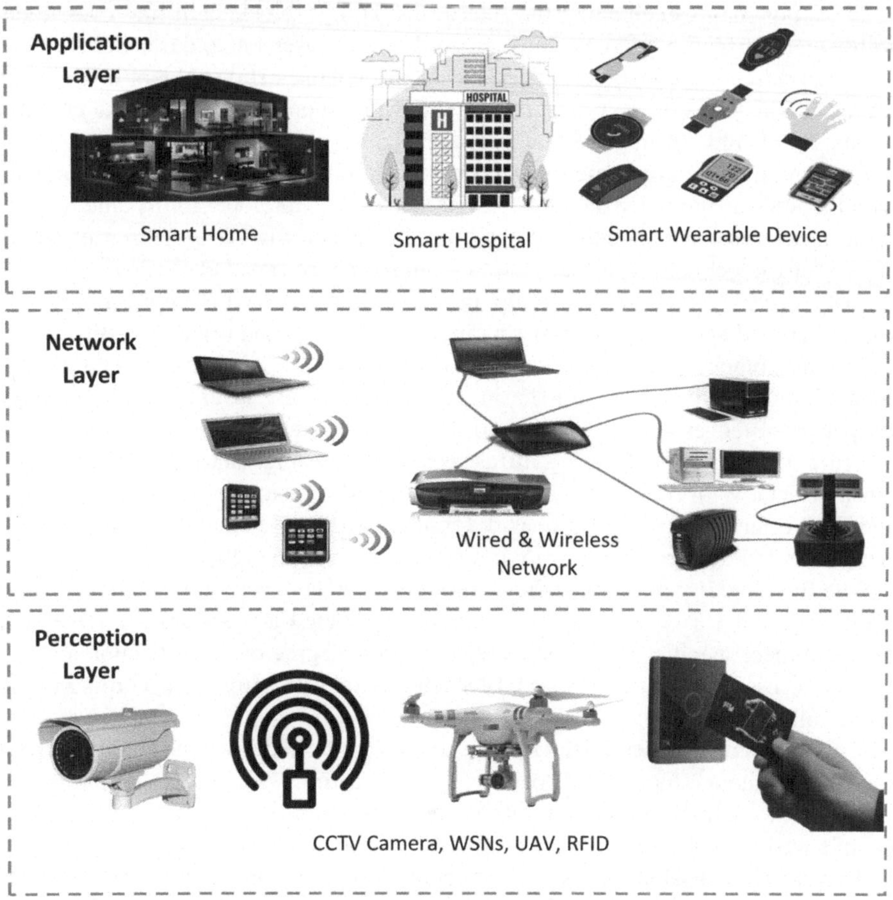


**FIGURE 3.1** Three-layered architecture of IoT.

In addition to security, privacy, and efficiency benefits, integrating blockchain with IoT can enable new business models. Blockchain enables the tokenization of assets, enabling IoT devices to exchange value and transact autonomously without the necessity of intermediaries. This creates more opportunities for new monetization models, such as micropayments for data sharing, peer-to-peer energy trading in smart grids, and automated leasing of IoT devices [16].

Blockchain enables interoperability and data sharing across heterogeneous IoT platforms and ecosystems. By facilitating a common and decentralized infrastructure for data exchange, blockchain facilitates error-free integration and collaboration among diverse IoT devices and platforms. This interoperability promotes an innovation method and increases the development of new IoT applications and services by enabling data sharing and collaboration across different industries and use cases [17].

In supply chains enabled by the Internet of Things, blockchain improves traceability and transparency. By keeping a record of the movement and transfer of goods on a transparent and absolute ledger, blockchain enables stakeholders to trace the source and histories of the product along the whole supply chain, reducing counterfeiting, fraud, and unauthorized tampering [18].

To sum up, there are numerous advantages to combining blockchain technology with IoT, such as improved security, privacy, efficiency, interoperability, and the creation of new business models. By applying the decentralized, tamper-proof ledger of blockchain technology, IoT ecosystems can solve traditional challenges related to security, privacy, and data ownership, making it possible for IoT applications to be widely adopted and deployed across a range of industries and use cases [19].

Overall, blockchain integration and IPFS provide a compelling solution for enhancing user privacy in data-sharing environments. By leveraging blockchain's tamper-proof ledger and IPFS's decentralized file storage, organizations can create a secure and privacy-enhancing infrastructure for storing, sharing, and accessing sensitive data. With concerns about data privacy and security on the rise, blockchain, and IPFS offer a promising framework for addressing these challenges and empowering users to retain control over their personal information [20].

The functionality of blockchain technology extends beyond the simple management of digital currencies. Several kinds of distributed ledgers are available, and each can meet specific requirements regarding the degree of centralization and the type of access control used [21–25]. Blockchains are distinguished according to the degree of centralization:

**Public (permissionless)**: Blockchains are not controlled since the consensus process behind them is open to all participants and entirely decentralized, and blockchains allow anybody to submit data to the distributed ledger. Public blockchains include well-known cryptocurrencies such as Bitcoin and Ethereum.

**Private (permissioned)**: An access control mechanism in blockchains determines who may participate in the network. Since the other nodes have verified every node, their identities are also known. There may be one or many owners of a permissioned blockchain, and only they have the authority to create and validate its data. Nodes that manage the construction of blocks might execute a consensus method or a set of specified nodes.

## PROBLEM STATEMENT

The hybridization of blockchain with IoT has massive opportunities for promoting different businesses by enabling secure, autonomous, and data exchange among interconnected devices. However, many challenges get delayed for the broad acceptance and successful application of blockchain-enabled IoT solutions.

- **Scalability**: The exponential growth of IoT devices created essential concerns about blockchain networks' scalability. It is difficult for traditional blockchain systems like Ethereum and Bitcoin to handle large numbers of transactions quickly and cost-effectively. For blockchain-enabled IoT applications, requiring real-time data processing and transaction validation to serve a variety of use cases like supply chain management, industrial automation, and smart cities, this scalability issue presents a major difficulty.

- **Security and privacy**: Although blockchain has inherent security advantages such as immutability and cryptographic algorithms, IoT devices are susceptible to cyberattacks due to their limited computing resources and lack of built-in security features. Securing IoT devices against malware is essential to preserving the security of data in blockchain-powered IoT network ecosystems. Additionally, preserving user privacy and ensuring compliance with data protection regulations remain important challenges, especially in applications involving sensitive personal or industrial data.
- **Interoperability**: The interoperability of blockchain networks and IoT devices is another critical challenge that controls or stops the various platforms and ecosystems. Increasing the number of diverse IoT devices targeted in achieving interoperability between blockchain networks and IoT devices is a difficult task. To deliver common protocols that promote interoperability and smooth integration of blockchain-enabled IoT solutions across various industries and use cases, standardization efforts are crucial.
- **Energy consumption**: Blockchain networks require a lot of techniques to secure the network and validate transactions, especially when using consensus methods based on proof of work (PoW). This high energy consumption is not sustainable, especially in IoT deployments where energy-efficient operation is crucial for battery-powered devices and remote sensors. Addressing the energy consumption of blockchain networks is essential to mitigate environmental impacts and ensure the feasibility of blockchain-enabled IoT solutions in resource-constrained environments.
- **Regulatory compliance**: Compliance with regulatory requirements and legal frameworks poses a significant challenge for blockchain-enabled IoT deployments, particularly in domains like supply chain management, healthcare, and finance. Regulatory uncertainties regarding data ownership, liability, and data protection laws obstruct the adoption of blockchain technology in IoT applications. Clarifying regulatory guidelines and establishing industry-specific standards are essential to address legal and compliance challenges and have blockchain-enabled IoT solutions.

In conclusion, while integrating blockchain technology with IoT has great potential to transform several industries, obstacles need to be overcome to fully achieve this potential. Scalability, security, privacy, interoperability, energy consumption, and regulatory compliance are among the key challenges that need to be overcome to facilitate the widespread adoption and effective implementation of blockchain-enabled IoT solutions. Technology developers, legislators, and industry stakeholders must come to work together to develop blockchain-enabled IoT systems that are interoperable, scalable, and secure to fulfill the changing needs of various industries and use cases.

## MOTIVATION

Blockchain technology and IoT integration have the capacity to completely transform several industries and tackle important privacy issues. The motivation behind exploring this convergence stems from several key factors:

- **Enhanced security**: Traditional centralized architectures for IoT data management are vulnerable to cyberattacks and prone to single points of failure. By leveraging the decentralized and tamper-proof ledger of blockchains, IoT ecosystems can enhance data security by guaranteeing the confidentiality and integrity of data transactions. The idea to integrate blockchain with IoT lies in modifying security risks and preserving stakeholder trust by offering an absolute and transparent record of transactions of devices.
- **Improved privacy**: With the proliferation of IoT devices collecting sensitive personal and industrial data, concerns about data privacy and ownership have become paramount. Blockchain technology enables users to control their data by providing secure and transparent access control mechanisms. The idea to integrate blockchain with IoT lies in empowering individuals and organizations to uphold data ownership.

**Seamless interoperability**: The interoperability of blockchain networks and IoT devices is important to making collaboration and data exchange seamless among heterogeneous platforms and ecosystems. To provide common frameworks and protocols, standardization initiatives are relevant that enable interoperability and seamless integration of blockchain-enabled IoT solutions across different industries and use cases. The idea to integrate blockchain with IoT lies in fostering innovation and accelerating the development of new IoT applications and services by enabling data sharing and collaboration across diverse platforms and ecosystems.

- **Autonomous transactions**: Through the use of smart contracts, blockchain provides safe, autonomous transactions between Internet of Things devices without the need for intermediaries. This removes the requirement for centralized authorities, streamlines processes, and reduces transaction costs associated with traditional payment systems. The idea to integrate blockchain with IoT lies in enabling new use cases, business models, and opportunities for innovation by enabling autonomous transactions and value exchange among interconnected devices.
- **Environmental sustainability**: To ensure the viability of blockchain-enabled IoT solutions, it is imperative to access energy requirements, especially in resource-constrained environments. The idea to integrate blockchain with IoT lies in developing energy-efficient blockchain consensus mechanisms and protocols that mitigate environmental impacts and enable sustainable IoT deployments.

In summary, the motivation behind exploring the integration of blockchain with IoT stems from the need to enhance security, privacy, interoperability, and autonomy in IoT ecosystems while addressing critical challenges and unlocking new opportunities for innovation in various industries and use cases.

## RESEARCH OBJECTIVES

- To investigate the feasibility and scalability of integrating blockchain technology with the IoT to conjure data security and privacy in diverse use cases, including smart cities, healthcare, and industrial automation.
- To assess the effectiveness of blockchain-enabled IoT solutions in mitigating security risks and ensuring data integrity, confidentiality, and availability in IoT ecosystems, with a focus on identifying vulnerabilities and potential attack vectors.
- To evaluate blockchain technology's impact on enhancing user privacy and data ownership in IoT deployments, including the implementation of decentralized access control mechanisms and adherence to laws governing data privacy, like the General Data Protection Regulation (GDPR).
- To explore strategies to improve blockchain and IoT devices interoperability, protocols, and platforms to facilitate collaboration across heterogeneous environments and ecosystems.
- To investigate energy-efficient blockchain consensus mechanisms and protocols suitable for IoT deployments to minimize energy consumption and environmental impacts while ensuring the reliability of blockchain-enabled IoT solutions.
- To assess the controlling system governing blockchain-enabled IoT deployments, including liability, and compliance with industry-specific regulations and standards.
- To develop and evaluate prototype blockchain-enabled IoT solutions for specific use cases, demonstrating the possibility, effectiveness, and feasible benefits of integrating IoT and blockchain.
- To identify challenges, limitations, and best practices associated with the integration of blockchain with IoT and offer suggestions for resolving these issues and optimizing the advantages of blockchain-enabled IoT systems.
- To explore opportunities for innovation and value creation enabled by blockchain-enabled IoT solutions, including new business models, revenue streams, and opportunities for collaboration and partnership across industries and sectors.
- To provide insights and recommendations for policymakers, industry stakeholders, and technology innovators on the adoption, implementation, and optimization of blockchain-enabled IoT solutions to address critical challenges and unlock new opportunities for innovation and growth.

## BLOCKCHAIN IN IOT

Blockchain's primary innovation is the ability to achieve consensus among participants without reliance on a central authority. Blockchain technology makes use of advanced cryptography to provide extra characteristics such as decentralization, immutability, consensus processes, and smart contracts. The primary focus of advanced cryptography is the protection of data and communications. The combination of cryptography and blockchain technology provides a higher level of security and trust, in addition to opening new opportunities for a wide range of applications.

Blockchain technology is being used for a variety of purposes, and although advanced encryption is an essential component, blockchain also provides a number of other benefits that are entirely unique. Poor network visibility in an IoT environment is the next security concern. As the number of devices in IoT is rising exponentially, providing generalized security for every device is becoming increasingly challenging.

Because of the many IoT devices and dynamic factors, traditional information technology security approaches require accounting for every component linked to the network, which is not practical in IoT. There is still a problem with the link between mobility and security. Figure 3.2 lists the many types of IoT security vulnerabilities. Another critical security vulnerability in the IoT context is ignorance among users. It would not be prudent to assume that users are aware of every possible method of attack and weak spots in the user environment. Users are less concerned about security vulnerabilities related to IoT device use and more focused on how convenient the devices are to use.

Another factor contributing to IoT security problems is the way an IoT component uses the communication channel. Currently, several protocols are employed in an IoT communication system that have already been the subject of security incidents and are known to adversely affect the whole networking system. Although there are many different kinds of assaults in an IoT context, some of the most frequent ones are as follows:

- **Physical tampering**: When devices are operated in an environment that is inaccessible to humans, there is a risk of unapproved tampering with the physical components of the device.
- **Denial-of-service**: Denial-of-service assaults are among the most common in cloud and IoT environments. They cause the entire network and servers to go down, making it simpler for intruders to get onto the system or carry out other harmful goals.
- **Firmware jacking**: If the system is attacked, firmware updates may also be affected. When a compromised firmware update gets into devices, it affects the devices as well because authentication of the firmware update's origination point is unsuccessful.
- **Injection of malicious node**: The attacker manually inserts the harmful nodes into the typical, everyday IoT gadgets. The malicious nodes are capable of illicit data sniffing and other types of unlawful network control.

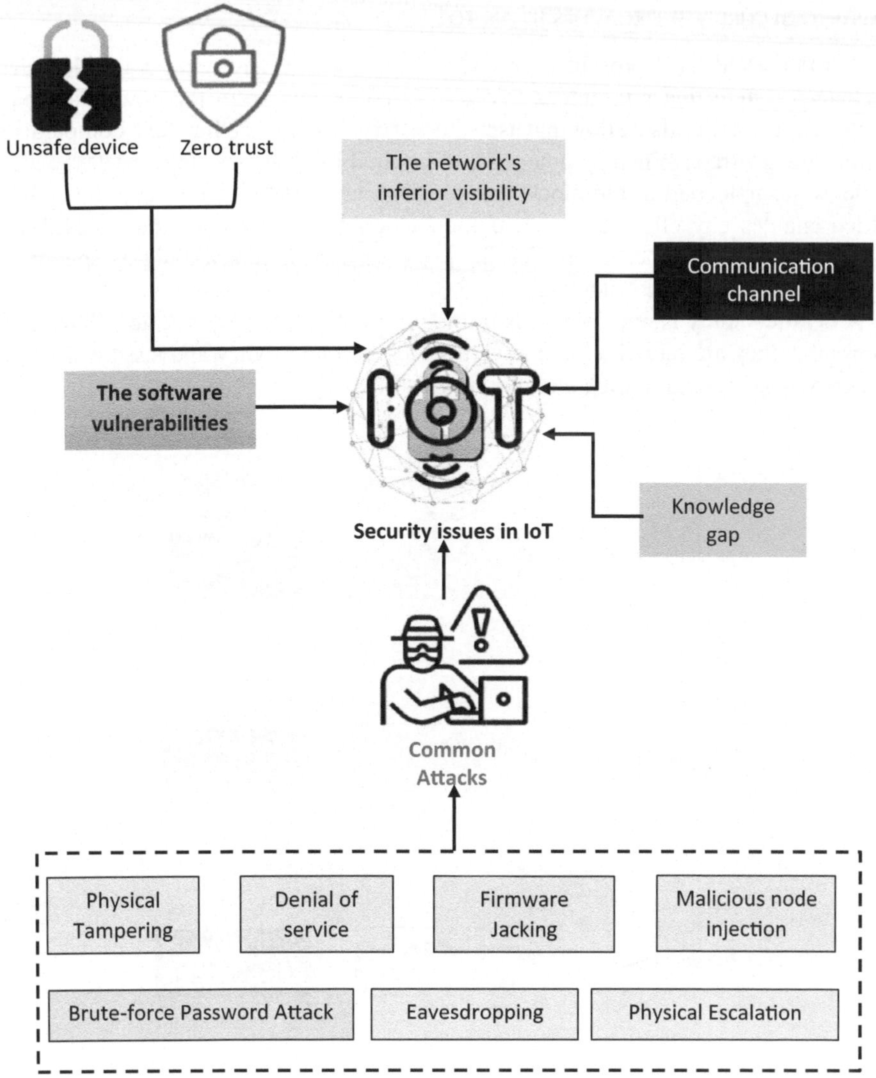


**FIGURE 3.2** IoT security issues.

- **Brute-force password attack**: In susceptible IoT systems, these kinds of assaults are also used to get access to devices.
- **Eavesdropping**: The attacker node can breach and intercept the weak communication channel. There are various weak points in the communication route where the attacker can obtain all the confidential information.
- **Physical escalation**: An attacker may use a variety of operating systems, hardware, and software weaknesses to gain unauthorized access to Internet of Things resources.

## Existing Security Approaches in an IoT

Al Hwaitat et al. [13] present a procedure for a secure and computable IoT data exchange system that guarantees that users may benefit from the computing ease of the cloud while also allowing users to verify that the cipher text computation results are accurate. When it comes to fine-granted IoT, data sharing and data access policies are uploaded to the blockchain through attribute-based encryption (ABE), which enables any client to check if their characteristics satisfy the accessibility policy from time to time. At present, the IoT security solutions are highly specific to layers of the IoT protocol stack (Figure 3.3).

A detailed study is required to examine the existing security solution. The result shows that they are more inclined toward strengthening privacy and less toward data integrity, service availability, and maintainability.

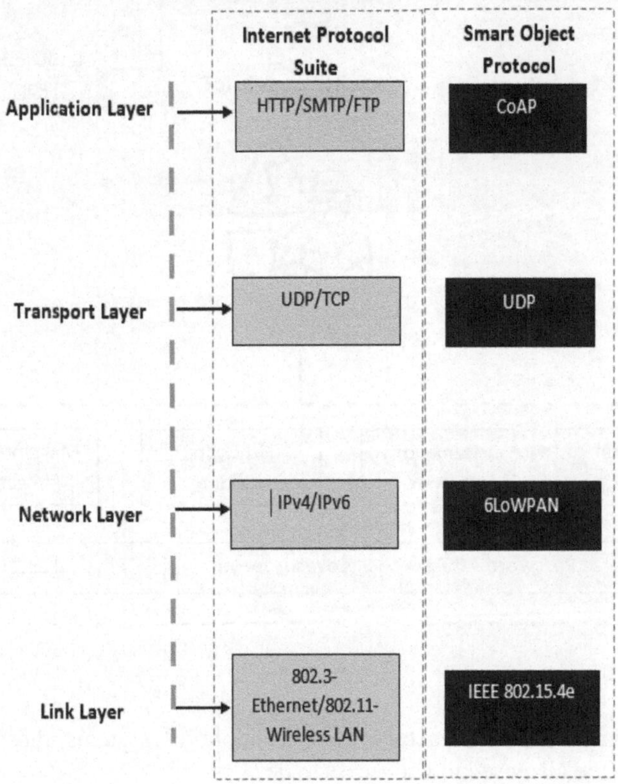


**FIGURE 3.3** IoT protocol suite.

## Adoption of Blockchain for Enhanced Security

At present, blockchain technology is seen as the long-term security approach in IoT for multiple reasons. Blockchain induces information bound in the form of a block and is connected using cryptography. A chain of blocks is made. Blockchain is characterized by integrity, security, and faster processing. However, there are variously reported limitations of blockchain, and the concept is still in the initial stage of development.

**Enhanced data security**: The decentralized nature of blockchain technology and the application of cryptographic techniques make it possible to store medical information in a way that is both secure and difficult to alter. Due to the distributed nature of blockchain design, it is exceedingly difficult for malicious actors to alter or manipulate the data.

**Interoperability and data exchange**: The use of blockchain technology provides a framework that is standardized for data interchange between the many different healthcare organizations, systems, and platforms. It makes seamless interoperability feasible, assures accurate and consistent data exchange across all parties, avoids unnecessary duplication, and improves the coordination of healthcare.

**Data integrity and immutability**: A chronological and permanent record of every transaction and input is kept by the blockchain, which results in an audit trail that cannot be altered in any way. This function ensures that the data included in medical records are accurate and prevents unauthorized changes from being made. As a result, the electronic health record (EHR) system becomes more trustworthy and reliable.

**Efficient and accurate auditing**: Due to the immutability and transparency of blockchain technology, auditing may be conducted more efficiently. The capacity for healthcare practitioners, authorities, and auditors to readily trace and verify the origin, access, and change histories of health information improves accountability and compliance.

**Disaster recovery and resilience**: The distributed design of a blockchain makes it feasible to implement disaster recovery and resilience strategies.

## Secure Blockchain-Based IoT Information Storage and Retrieval

Blockchain suffers from space constraints when storing large files as every node maintains complete network data. IPFS provides a promising solution to the present issue. In IPFS, the hashed value of the raw data is embedded in the chain transaction. Unless private information is stored, IPFS doesn't permit preselected nodes to store the data. IPFS uses an access control list for distributing hashed data among listed communicating parties. Smart contracts are employed in the Ethereum blockchain to provide the access control policy to the requester. Smart contracts are verified before file uploading and retrieving.

The IPFS is a peer-to-peer network protocol for storing data among obtainable distributed file systems. The content addressing scheme uniquely identifies each file in this distributed system. The content is stored as a distributed hash table (DHT) and any node upon request can retrieve the content data in the form of a hash. The data is uploaded as a 256 kilobyte file that is hashed further and identified with a content identifier. Without a server connection, the host can easily download the hash content from its peer nodes. For efficient storing and retrieving of a file, the full file is connected with a Merkle directed acyclic graph (DAG). The DAG additionally helps to connect a node with its specific identifier. Therefore a node uses DHT to retrieve IPFS data.

Figure 3.4 has three layers: perception layer, blockchain network layer, and application layer. The crude drug information created by a QR code scanner at the collecting site is provided by the perception layer.

Generation of raw pharmaceutical data (D) at the manufacturing site is done using a QR code scanner. AES is faster in encrypting both small and large files as compared to RSA. The AES key distribution is troublesome. Hence AES is used to encrypt the data and RSA is used to transfer encrypted AES key over the communication. Symmetric key K is encrypted with the public key of the retailer (R), supplier (S), consumer (C), admin (Ad), data manager (DM), and patient (P) to produce the cipher key in Equations 3.1–3.8.

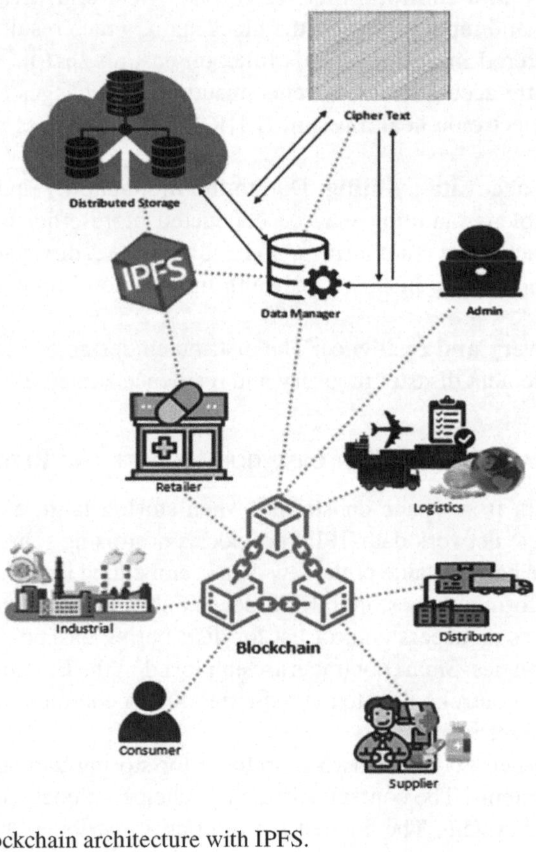


**FIGURE 3.4** Blockchain architecture with IPFS.

$$CK_R = RSA(K, PK_R) \tag{3.1}$$

$$CK_S = RSA(K, PK_s) \tag{3.2}$$

$$CK_D = RSA(K, PK_D) \tag{3.3}$$

$$CK_C = RSA(K, PK_C) \tag{3.4}$$

$$CK_{Ad} = RSA(K, PK_{Ad}) \tag{3.5}$$

$$CK_{DM} = RSA(K, PK_{DM}) \tag{3.6}$$

$$CK_P = RSA(K, PK_P) \tag{3.7}$$

$$CK_R = RSA(K, PK_R) \tag{3.8}$$

$$k = Decrypt - RSA\left(PR_P\left(CK_P\right)\right) \tag{3.9}$$

$$Original\,data(D) = Decrypt - AES\left(K\left(CD\right)\right) \tag{3.10}$$

## Design of Smart Contracts

The smart contract is generated by the data manager node.

Figure 3.5 describes the smart contract function executed in terms of transaction and execution costs for encryption, digital signature, uploading, and sending data to IPFS in the IPFS smart contract.

Figure 3.6 describes the smart contract function executed in terms of transaction and execution cost for node deployment, node registration, data request, and data response operations in the blockchain node deployment smart contract.

Figure 3.7 describes the smart contract function executed in terms of transaction and execution cost for user enrollment, user exit, applying regulation scheme, reclamation of user data, and user rejection functions in the smart contract access control policy scheme.

## Challenges and Issues

**Scalability**: The integration of blockchain with the IoT poses scalability challenges, particularly in handling the increasing volume of transactions generated by IoT devices. Traditional blockchain platforms, such as Bitcoin and Ethereum, may struggle to process a large number of transactions in a timely and cost-effective manner, leading to delays and increased transaction fees.

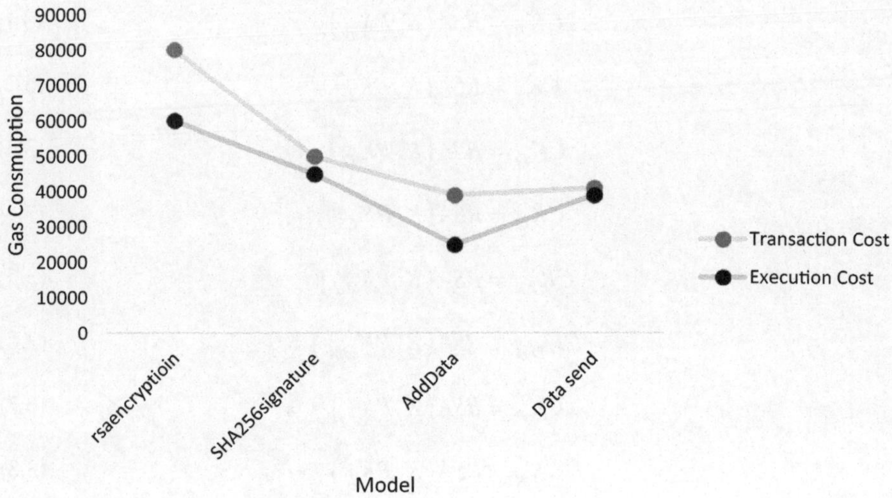


**FIGURE 3.5** Smart contract IPFS.

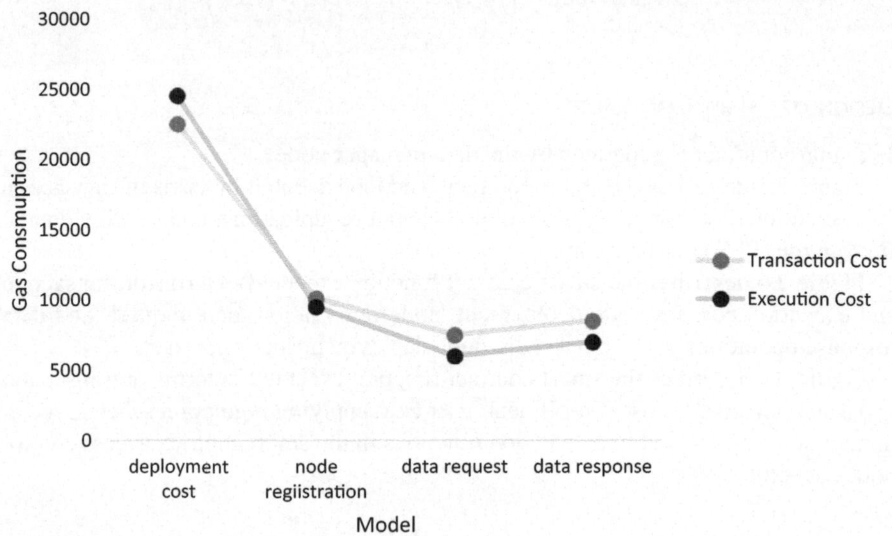


**FIGURE 3.6** Smart contract blockchain node function.

**Latency**: Delays in transaction confirmation may impact the responsiveness of IoT devices, particularly in applications that require low-latency communication, such as industrial automation and autonomous vehicles.

**Energy consumption**: Blockchain networks need to validate transactions and secure the network. This high energy consumption is not sustainable, particularly

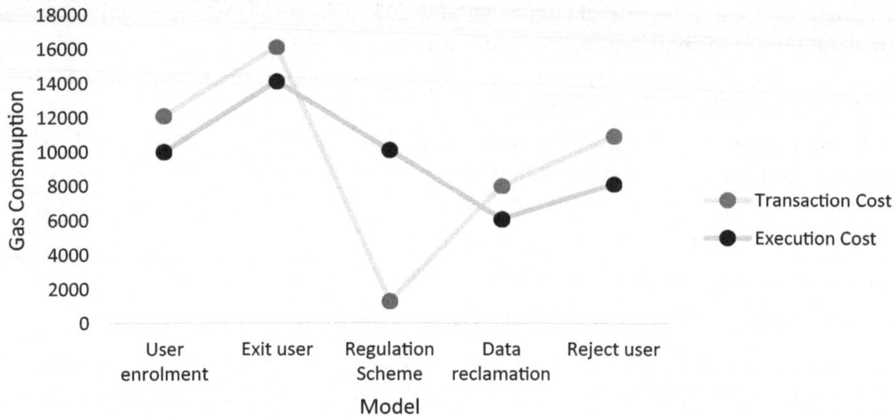


**FIGURE 3.7** Smart contract access control policy function.

for IoT deployments where energy-efficient operation is crucial for battery-powered devices and remote sensors.

**Interoperability**: Due to the proliferation of diverse IoT devices in attaining interoperability between blockchain networks and IoT devices has become challenging in maintaining protocols and standards. Lack of standardization and compatibility issues hinder seamless integration and data exchange among heterogeneous platforms and ecosystems, limiting the scalability and adoption of blockchain-enabled IoT solutions.

**Security risks**: While blockchain offers integral security benefits such as immutability and cryptographic algorithms, IoT devices are often vulnerable to cyberattacks due to their limited computing resources and lack of built-in security features. Securing IoT devices against malware, data breaches, and unauthorized access is crucial to maintaining the integrity and confidentiality of data in blockchain-enabled IoT ecosystems.

**Privacy concerns**: Preserving user privacy and ensuring compliance with data protection regulations remain significant challenges in blockchain-enabled IoT deployments. Blockchain's transparent and immutable ledger may expose sensitive information raising concerns

**Regulatory compliance**: Compliance with regulatory requirements and legal frameworks poses challenges for blockchain-enabled IoT deployments, particularly in industries such as healthcare, finance, and supply chain management. Regulatory uncertainties regarding data ownership, liability, and data protection laws hinder the adoption of blockchain technology in IoT applications.

**Complexity**: Integrating blockchain with IoT introduces complexity in terms of managing and maintaining the infrastructure, deploying smart contracts, and ensuring interoperability among different platforms and ecosystems. The complexity of blockchain-enabled IoT solutions may deter adoption, particularly among organizations with limited technical expertise and resources.

**Cost**: Implementing blockchain-enabled IoT solutions involves upfront costs for deploying and maintaining the infrastructure, developing smart contracts, and ensuring compliance with regulatory requirements. High implementation costs may hinder the adoption of new technology.

**Legacy systems integration**: Integrating blockchain with existing IoT infrastructure and legacy systems may lack support for blockchain technology, requiring significant modifications or upgrades to enable seamless integration with blockchain-enabled IoT solutions.

## CONCLUSION AND FUTURE ENHANCEMENT

The integration of blockchain with IoT holds immense potential and significant progress has been made in exploring and implementing solutions to overcome these obstacles. Blockchain technology offers inherent security benefits with cryptographic algorithms, which enhance the integrity and confidentiality of data in IoT ecosystems. By leveraging blockchain's decentralized architecture and smart contract capabilities, IoT devices can securely record transactions, execute autonomous transactions, and ensure data privacy and ownership.

Moreover, blockchain enables seamless interoperability and data sharing across heterogeneous IoT platforms and ecosystems, fostering collaboration and innovation in diverse industries and use cases. Standardization efforts and industry collaborations are essential to develop common protocols and enable seamless integration of blockchain-enabled IoT solutions. Despite the progress made in addressing challenges and issues, several areas for future enhancement and research remain:

**Scalability**: Scalable blockchain solutions have to handle a large quantum of transactions generated by IoT devices. This involves exploring alternative consensus mechanisms to improve throughput and reduce latency in blockchain networks.

**Energy efficiency**: Addressing the energy consumption is crucial to ensure the feasibility of blockchain-enabled IoT solutions, particularly in resource-constrained environments. The recommendations for future researchers are protocols that minimize energy consumption while ensuring the scalability and reliability of blockchain networks.

**Privacy-preserving technologies**: Future research should explore privacy-preserving technologies and techniques that enable secure and anonymous data sharing in blockchain-enabled IoT ecosystems. This involves developing techniques to protect data.

**Interoperability**: Further efforts are needed to enhance interoperability between blockchain networks and IoT devices, protocols, and platforms. The recommendations for future research include developing standard protocols, APIs, and interoperability frameworks that enable seamless data exchange and collaboration among heterogeneous IoT environments.

**Regulatory compliance**: Addressing regulatory challenges and legal frameworks governing blockchain-enabled IoT deployments requires collaborative efforts from policymakers, industry stakeholders, and technology innovators. Future studies should clarify regulatory guidelines, establish industry-specific standards, and foster collaboration to ensure compliance with data protection regulations and legal requirements.

**User experience**: Enhancing the user experience of IoT-enabled blockchain solutions is essential to drive adoption and usability among diverse stakeholders. The recommendations for future research are to develop user-friendly interfaces, tools, and educational resources that empower users to understand and interact with blockchain-enabled IoT applications effectively.

In conclusion, while significant progress has been made in exploring and implementing blockchain-enabled IoT solutions, there remains opportunities for future enhancement and research to address scalability, energy efficiency, and user experience challenges. By addressing these areas, blockchain technology can use IoT, enabling security, transparency, and collaboration across diverse industries and use cases.

## REFERENCES

1. Tyagi, A. K., Dananjayan, S., Agarwal, D., & Thariq Ahmed, H. F. (2023). Blockchain—Internet of Things Applications: Opportunities and Challenges for Industry 4.0 and Society 5.0. *Sensors*, *23*(2), 947.
2. Issa, W., Moustafa, N., Turnbull, B., Sohrabi, N., & Tari, Z. (2023). Blockchain-based federated learning for securing internet of things: A comprehensive survey. *ACM Computing Surveys*, *55*(9), 1–43.
3. Latchoumi, T. P., Swathi, R., Vidyasri, P., & Balamurugan, K. (2022). Develop new algorithm to improve safety on WMSN in health disease monitoring. In *2022 International Mobile and Embedded Technology Conference (MECON)* (pp. 357–362). IEEE.
4. Balamurugan, K., Latchoumi, T. P., & Ezhilarasi, T. P. (2022). Wearables to improve efficiency, productivity, and safety of operations. In *Smart Manufacturing Technologies for Industry 4.0* (pp. 75–90). CRC Press.
5. Abed, S. E., Jaffal, R., & Mohd, B. J. (2023). A review on blockchain and iot integration from energy, security and hardware perspectives. *Wireless Personal Communications*, *129*(3), 2079–2122.
6. Hu, H., Xu, J., Liu, M., & Lim, M. K. (2023). Vaccine supply chain management: An intelligent system utilizing blockchain, IoT and machine learning. *Journal of Business Research*, *156*, 113480.
7. Latchoumi, T. P., Parthiban, L., Balamurugan, K., Raja, K., Vijayaraj, J., & Parthiban, R. (2023). A framework for low energy application devices using blockchain-enabled IoT in WSNs. In *Integrating Blockchain and Artificial Intelligence for Industry 4.0 Innovations* (pp. 121–132). Springer International Publishing.
8. Yazdinejad, A., Dehghantanha, A., Parizi, R. M., Srivastava, G., & Karimipour, H. (2023). Secure intelligent fuzzy blockchain framework: Effective threat detection in IoT networks. *Computers in Industry*, *144*, 103801.
9. Cai, J., Liang, W., Li, X., Li, K., Gui, Z., & Khan, M. K. (2023). GTxChain: A secure IoT smart blockchain architecture based on graph neural network. *IEEE Internet of Things Journal*, *10*(24), 21502–21514.
10. Aroul Anandam, V. V., Latchoumi, T.P., Balamurugan, K., & Yookesh, T. L. (2020). Improving the energy efficiency in mobile ad-hoc network using learning-based routing. *Revue d'Intelligence Artificielle*, *34*(3), 337–343.

11. Arun karthikeyan, K., & Balamurugan, K. (2020). Performance improvement of Cryo treated insert on turning studies of AISI 1018 steel using multi objective optimization. In *2020 International Conference on Computational Intelligence for Smart Power System and Sustainable Energy (CISPSSE)* (pp. 1–4). IEEE.
12. Kumar, D., Singh, R. K., Mishra, R., & Daim, T. U. (2023). Roadmap for integrating blockchain with Internet of Things (IoT) for sustainable and secured operations in logistics and supply chains: Decision making framework with case illustration. *Technological Forecasting and Social Change, 196*, 122837.
13. Al Hwaitat, A. K., Almaiah, M. A., Ali, A., Al-Otaibi, S., Shishakly, R., Lutfi, A., & Alrawad, M. (2023). A new blockchain-based authentication framework for secure IoT networks. *Electronics, 12*(17), 3618.
14. Sharma, P., Namasudra, S., Crespo, R. G., Parra-Fuente, J., & Trivedi, M. C. (2023). EHDHE: Enhancing security of healthcare documents in IoT-enabled digital healthcare ecosystems using blockchain. *Information Sciences, 629*, 703–718.
15. Taloba, A. I., Elhadad, A., Rayan, A., Abd El-Aziz, R. M., Salem, M., Alzahrani, A. A., ... & Park, C. (2023). A blockchain-based hybrid platform for multimedia data processing in IoT-Healthcare. *Alexandria Engineering Journal, 65*, 263–274.
16. Kumar, P., Kumar, R., Gupta, G. P., Tripathi, R., Jolfaei, A., & Islam, A. N. (2023). A blockchain-orchestrated deep learning approach for secure data transmission in IoT-enabled healthcare system. *Journal of Parallel and Distributed Computing, 172*, 69–83.
17. Oudani, M., Sebbar, A., Zkik, K., El Harraki, I., & Belhadi, A. (2023). Green blockchain based IoT for secured supply chain of hazardous materials. *Computers & Industrial Engineering, 175*, 108814.
18. Zubaydi, H. D., Varga, P., & Molnár, S. (2023). Leveraging blockchain technology for ensuring security and privacy aspects in internet of things: A systematic literature review. *Sensors, 23*(2), 788.
19. Koe, A. S. V., Ai, S., Chen, Q., Tang, J., Chen, K., Zhang, S., & Li, X. (2023). Hieraledger: Towards malicious gateways in appendable-block blockchain constructions for IoT. *Information Sciences, 632*, 87–104.
20. Khan, A. A., Shaikh, A. A., & Laghari, A. A. (2023). IoT with multimedia investigation: A secure process of digital forensics chain-of-custody using blockchain hyperledger sawtooth. *Arabian Journal for Science and Engineering, 48*(8), 10173–10188.
21. Sammeta, N., & Parthiban, L. (2022). Hyperledger blockchain enabled secure medical record management with deep learning-based diagnosis model. *Complex & Intelligent Systems, 8*(1), 625–640.
22. Sammeta, N., & Parthiban L. (2021). Medical data analytics for secure multi-party-primarily based cloud computing utilizing homomorphic encryption. *Journal of Scientific and Industrial Research (JSIR), 80*, 692–698.
23. Sammeta, N., & Parthiban, L. (2022). An optimal elliptic curve cryptography based encryption algorithm for blockchain-enabled medical image transmission. *Journal of Intelligent & Fuzzy Systems, 43*(6), 8275–828.
24. Sammeta, N., & Parthiban, L. (2023). Blockchain-based scalable and secure EHR data sharing using proxy re-encryption. *International Arab Journal of Information Technology*, 20(5), 702–710.
25. Sammeta, N., & Parthiban, L. (2021). Data ownership and secure medical data transmission using optimal multiple key-based homomorphic encryption with hyperledger blockchain. *International Journal of Image and Graphics, 23*(3), 2240003.

# *Section–II*

## *Industrial and societal implications using blockchain and IoT*

# 4 Exploration and Review of Blockchain Technology Used in Industrial Solutions

*Maxraj J and VijayAnanth Suyamburajan*

## INTRODUCTION

Following a paper titled "A Peer-to-Peer Electronic Cash System," published in 2008 by someone under the pseudonym Satoshi Nakamoto, blockchain technology gained widespread public awareness with the introduction and use of the Bitcoin network. The blueprint schemes described in the article are used by many contemporary cryptocurrencies. The first of numerous blockchain-based apps is Bitcoin.

The various elements of blockchain technology and their key characteristics that make blockchain a compelling technology can sometimes make it challenging to understand. Blockchain is in the form of blocks that create distributed digital ledgers that hold transfers that have been signed securely in each block. Further, a block itself is cryptographically linked to the previous block forming a linear chain of blocks, after validations and undergoing consensus decisions. Because of this, if the data is altered in one block, its unique identifier will change and be visible in the next block (providing tamper evidence). All users within the blockchain can determine whether the data in a prior block has been altered with this domino effect. A blockchain network offers a robust way of cooperative recordkeeping because it is challenging to alter or delete.

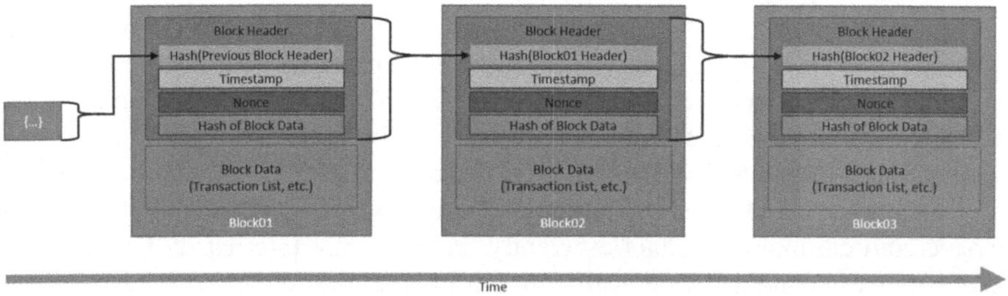


DOI: 10.1201/9781003428886-6

Blockchain technology at its core combines three mathematical constructs—cryptography, distributed systems, and game theory—in a meaningful and elegant way. Blockchain technology has the ability to be integrated into a wide range of systems, including data registers, digital identification, manufacturing supply chains, and records management.

## BLOCKCHAIN TECHNOLOGY FOR INDUSTRIES AND ENTERPRISES

The primary focus of blockchain is on financial and legal transactions that take advantage of the benefits of decentralized databases and a peer-to-peer approach beyond corporate borders. Enterprise entities that are interested in adopting blockchain typically go through trial and discovery phases before deciding on the most practical and applicable use cases. In this chapter, the blockchain technology adapted by Siemens is discussed in detail. Blockchain is significant for Siemens as a technology leader for providing services on a digital basis. Blockchain is therefore deemed one of the core technologies in Siemens and researched for its various applications and use cases.

This chapter examines a few of the applications of permissioned blockchain in products and services offered by Siemens companies and industries, including:

1. Digitalizing the supply chain for the food and beverage industry.
2. Distributed and transparent management of electricity and energy distribution.
3. Seamless transaction of untrusted parties in car sharing.
4. Supply chain management of tools in servicing operations of power plants.

## DIGITALIZING THE SUPPLY CHAIN FOR FOOD AND BEVERAGE INDUSTRY

As per WHO food safety facts [1], every year, one in ten individuals around the world gets sick after eating contaminated food. A study from NetNames [2] claims that counterfeiting is increasing by 15% yearly and depleting more than 2% of the world's GDP. It also emphasizes that consumers avoid companies for two years on average after a negative experience and that they share negative experiences 45% of the time on social media. Additionally, 61% of consumers worldwide say they make an effort to spend more money on goods from companies that behave ethically.

The introduction of blockchain technology has already had an impact on the financial sector. It is time to understand and handle problems that the food and beverage industry has long faced, like food recalls due to contamination, counterfeiting, and catering to changing consumer preferences. The feasibility of the technology was previously threatened by the challenges associated with the collection and transfer of data to the blockchain in commercial contexts. It is now feasible to generate, manage, and communicate data efficiently thanks to the Internet of Things (IoT), cloud computing, and the wide availability of internet access.

MindSphere is a very popular cloud-based platform from Siemens that offers a way to collect data from supply networks. On MindSphere, there are prebuilt blockchain apps that make it simple to subscribe and participate. Companies can also restrict users' access to information using these apps. This allows for the privacy of confidential information while only sharing essential information with other supply chain participants. Secure cloud gateways are used to package and send data from IoT sensors, shop floor systems, and tools to the blockchain. Siemens is busy creating IoT sensors and modules for its current software in order to send data straight to the blockchain.

The Boston Consulting Group (BCG) analysis report [3] compared an IoT solution with blockchain to an IoT solution without blockchain, and it came to the conclusion that the solution with blockchain could "achieve net savings equal to 0.6% of revenues" on logistics and storage alone. Real-time traceability and smart contracts, which enable better inventory management, brokerage fees, fraud protection, and other advantages, are largely responsible for these advantages. A complete traceability document can be created as supply chain blind spots become clearer. This document can speed up root-cause analysis and shorten recall times. Brand owners and consumers can verify the genuineness of products or other central statements using data from the blockchain by utilizing custom rules and smart contracts.

## DISTRIBUTED AND TRANSPARENT MANAGEMENT OF ELECTRICITY AND ENERGY DISTRIBUTION

Automated pay-per-use, created by Siemens Energy, enables utilities to pay for upgraded services as they are used in exchange for a portion of the price when revenue is produced. (i.e., electricity is successfully sold). Blockchain makes sure that the power plant's precise profit is computed and distributed proportionally. This translates into the ability to shift service products that have high capital expenditure (CAPEX) costs for businesses into operating expenses (OPEX), leading to new financing models for Siemens Energy's goods and services. It is no longer required to amortize all costs at once, only the costs of installation when necessary and the costs associated with each use. Therefore, it is feasible to produce energy while accurately estimating and paying the upfront and ongoing costs. Customers and solution providers will eventually share some of the operating risks, according to such a plan. And this is where the paradigm truly makes a difference.

In Spain, Siemens Energy has introduced e-ing3ni@, an original blockchain platform for trading renewable energy that enables customers to choose and reserve electricity from their preferred energy sources. The network, the first of its kind, uses the security and traceability offered by blockchain to enable online data management, report access, and the purchase and sale of renewable energy by producers, consumers, and retailers. Siemens Energy is also working on new business models based on blockchain for secure storage of digital identities, automated ordering, and 3D printing of spare parts (e.g., turbine parts) from suppliers.

In order to provide microgrid control solutions that maximize control over energy usage, Siemens is also utilizing blockchain technology. This is not Siemens's first foray into the world of microgrid development. In 2016, the firm teamed up with US startup LO3 Energy. With the help of this new effort, it is possible to create local trading between energy producers and consumers.

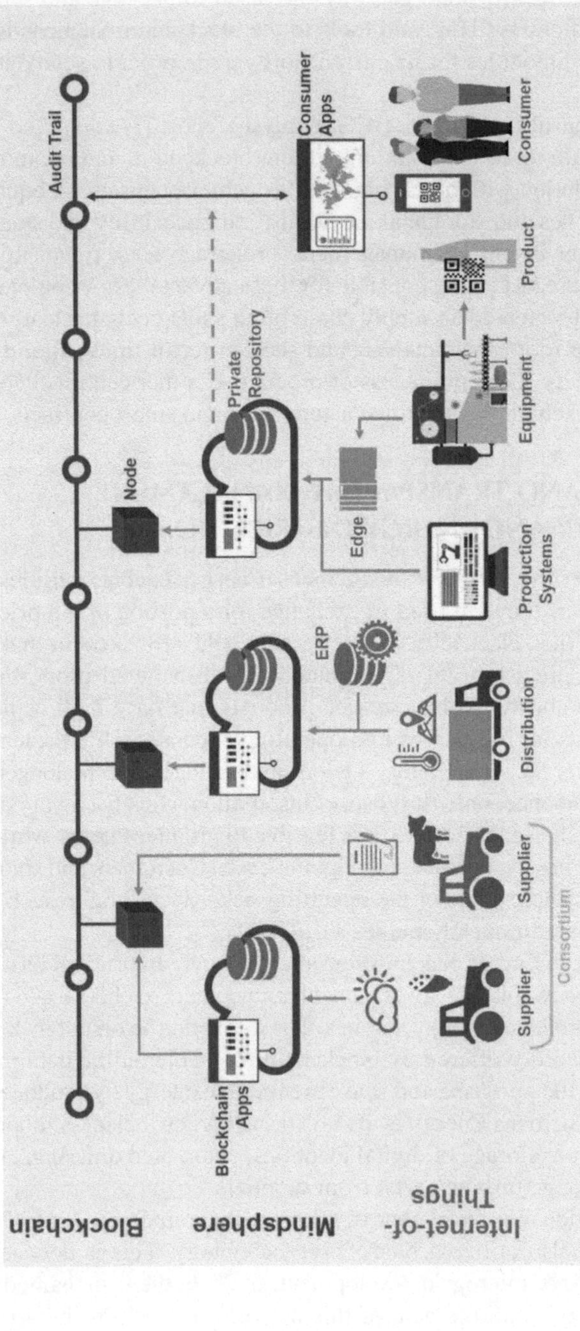


**FIGURE 4.1** Transparency and traceability with Siemens blockchain.

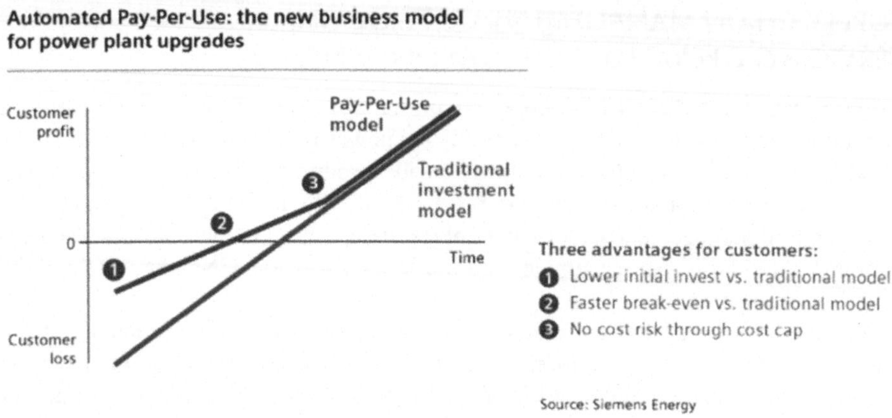


**FIGURE 4.2** Automated pay-per-use with Siemens blockchain.

## SEAMLESS TRANSACTION OF UNTRUSTED PARTIES IN CAR SHARING

There will be an increasing trend in the opportunities for mobility services such as car sharing rather than owning a car. People want to travel from one location to another, rather than wanting to own a car and it is much more convenient sometimes than owning a vehicle. Siemens Mobility focuses on providing transport solutions related to the movement of people and goods to various locations globally.

People who reside or work in large urban areas may frequently depend on car sharing to avoid some of the responsibilities associated with owning a car, such as finding parking, car payments, and insurance. According to information on the Enterprise CarShare website 4], car sharing enables users to use or rent a car for any duration, for, say, a minimum of one hour. Another illustration of a car sharing business is Zipcar [5].

Car sharing has its own challenges. Though the service ensures that one doesn't have to pay for parking, fueling, etc., customers are often faced with the challenge of finding specific parking and fueling stations that are in collaboration with the respective car sharing company. Another challenge for the car sharing companies is managing the fueling card. Once the customer fuels the car through authorized fueling stations, they must use a fuel card and it is associated with a PIN. It's not only inconvenient to customers but it's inconvenient to car sharing companies to manage these cards because the fueling cards get lost/stolen or are fraudulently sold on the internet.

Siemens Mobility is actively working on a blockchain solution to address these challenges and enable frictionless and secure transactions. Blockchain seamlessly integrates the interacting parties such as car companies, garage providers, fueling stations, and customers.

At Bosch's 2019 Connected World conference, the Siemens Corporate Technology team detailed the concept of "blockchain-based smart parking" as one possibility of blockchain usage.

## SUPPLY CHAIN MANAGEMENT OF TOOLS IN SERVICING OPERATIONS OF POWER PLANTS

Siemens uses blockchain in its supply chain management activities specifically for managing the tools in servicing operations for power plants all over the world. The SCM DigiNetwork, an informal company-wide group of digitalization enthusiasts who design and implement solutions in this field, was behind the original initiative. The tool management sector where blockchain is used involves a wide range of equipment and tools, from a simple cordless screwdriver to a CNC machine costing over a million euros.

The challenge that the tool management sector faces is that the contents of the return containers and the condition of the individual tools are very different from the shipments dispatched to the location of use in the first place. Previously, the number and condition of the devices and tools that were packed, transported, and received were recorded manually by the relevant employees. However, this manual documentation is error-prone and does not work very well. This is precisely why Siemens decided to use blockchain technology in a bid to guarantee the integrity of the newly digitalized process.

When a transport order is generated, it is automatically entered into the blockchain. This will show a list of associated and relevant partners (such as the logistics company), access to the document, and all papers associated with the freight shipment. All amendments are recorded and visible to all authorized parties. The blockchain technology used by the Siemens tool management sector gives Siemens the ability to, for example, view anytime whether a tool is still in customs or being transported, know whether the returned toolkits contain drills that need to be sharpened, or if an open-ended spanner is bent out of shape.

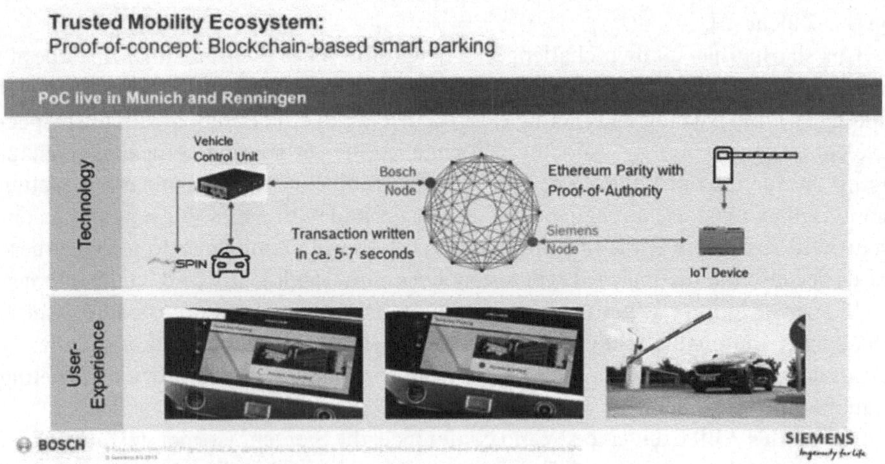


FIGURE 4.3 Blockchain-based smart parking solution from Siemens.

According to Maersk, the logistics company, administration costs can be higher than the physical transport costs for a short journey involving a container. However, they can be significantly reduced by using blockchain. Like Siemens, Maersk is also working on a blockchain project with its information technology supplier IBM.

## CONCLUSION

Blockchain has been a hot topic for several years. It has long been apparent that the distributed database and record system, first used for Bitcoin, is capable of much more than just enabling cryptocurrencies. Today, blockchain is used to build smart contracts that have variables and values specific to each transaction written directly into the code. It accomplishes this without the use of an intermediary, which is typically used to mediate real or electronic contracts between clients and customers.

The fact that blockchain is a decentralized system, which promotes trust in a digital ecosystem without the need for middlemen, as well as its transparency, which enables all users to audit transactions and precisely trace stored data back along the chain to its point of origin, are just a few of the key benefits that make it appealing for a variety of industries. The potential for creating new company models is enormous.

Due to the fact that blockchain creates confidence between parties who might not typically get along, future developments will focus more on shared goals and collaboration. Where there was no openness before, blockchain creates it.

Finally, blockchain technology is progressing as many organizations, like Siemens, are researching various business use cases.

## REFERENCES

1. WHO, 2017. "Food Safety." https://www.who.int/news-room/fact-sheets/detail/food-safety.
2. "NetNames report on Counting the Cost of Counterfeiting," n.d. https://www.yumpu.com/en/document/read/57066883/counting-the-cost-of-counterfeiting-a-netnames-report.
3. Siemens, n.d. "Blockchain IoT." https://xcelerator.siemens.com/global/en/industries/food-beverage/exclusive-area/blockchain-iot.html.
4. Enterprise, n.d. "FAQ." https://www.enterprisecarshare.com/us/en/faq.html.
5. Zipcar, n.d. "Zipcar," https://www.zipcar.com/.

# 5 Harnessing IoT for Effective Smart Farming

## *A Comparative Analysis of Farming Condition Monitoring*

*Anwer Basha H*

## INTRODUCTION

In the future, it is predicted that the influence of the Internet of Things (IoT) will serve as the primary foundation upon which intelligent computing will be built. The transition from "traditional technology" to "next generation everywhere computing" is an important endeavor that must be undertaken by both private and commercial sectors. The idea of an "Internet of Things" has recently emerged as one of the most significant topics of conversation in research groups all around the globe, with a special emphasis placed on advanced wireless communication technology. IoT is presently having a significant influence on people in a broad variety of settings all around the world. IoT is providing the groundwork, from the perspective of the average user, for the creation of a wide variety of goods, such as intelligent automation, intelligent living solutions, and intelligent healthcare services. In addition, it is often used for business reasons across a wide variety of industries, including manufacturing, transportation, agriculture, and corporate management, to name a few.

IoT has received the greatest attention and study in the realm of agriculture more than any other industry. The capacity of this business to provide food security is one of its most important aspects. This ability is becoming more important as the world's population continues to grow at an alarming rate. Researchers in this discipline first used methods that were based on ICT (information and communications technology). The difficulties faced by farmers are not solved by these strategies and are not a long-term answer to the problem. Researchers are investigating the possibilities of IoT technology in the agriculture industry as a feasible alternative to ICT. There are many different applications required for agricultural goods, such as managing supply chains, maintaining infrastructure, and monitoring soil moisture, temperature, and moisture levels in the surrounding environment. The use of IoT technology within the realm of agriculture has resulted in a new agricultural management paradigm

 DOI: 10.1201/9781003428886-7

referred to as "smart farming." This strategy seeks to maximize the amount of produce that can be harvested from agricultural land by using a wide range of technological instruments. Farmers have learned techniques in order to improve both the quality and quantity of their agricultural products. Smart farming practices require farmers to make use of many resources, including fertilizers. It has been noticed that agricultural laborers are not able to sustain a continuous presence in the field. In addition, it is conceivable that farmers do not possess the necessary expertise to assess the ideal environmental conditions for their crops via the use of a variety of technical instruments, and this is a possibility that should be taken into consideration. Farmers are given access to an IoT-based automated system that may function on its own and alert them to any possible issues that may arise in connection with farming. Because the technology is able to warn farmers even when they are not physically present in the field, it enables farmers to more effectively manage many farms and increase their agricultural production.

By the year 2050, the world population is anticipated to reach 9 billion people. In order to maximize the use of available resources and keep up with the requirements of a fast growing population, it is essential for the agricultural industry to make use of applications of IoT. The effects of global warming on farmers have resulted in significant losses owing to the increased frequency and severity of extreme weather events. However, the deployment of the IoT smart farming application provides a method to offset these losses by allowing farmers to rapidly adapt to changing weather conditions. This enables farmers to save more money. Researchers have provided a quantum of analysis of the relevance of smart farming as well as the possible future uses of IoT.

## LITERATURE REVIEW

IoT makes use of mobile devices such as smartphones to obtain data wirelessly from other mobile devices and to monitor the activity of living organisms such as plants and animals. Farmers are able to estimate output levels and evaluate weather conditions with the widespread use of sensors and gadgets in their operations. In the realms of water harvesting, flow monitoring and management, crop water needs assessment, supply scheduling, and water conservation, IoT is finding an increasing amount of use. Through the usage of sensors and cloud connection that is provided by a gateway, remote monitoring of the water supply and status can be carried out. This monitoring is performed in accordance with the particular demands that are placed on the soil and plants. It is not possible for farmers to do manual monitoring of each plant in order to treat nutrient deficits, eliminate pests, or prevent illnesses. In spite of this, the use of IoT technology has been shown to be helpful, and it has provided farmers with the opportunity to attain a new level of success in contemporary agriculture.

The expansion of IoT technology in recent years has brought about substantial advantages for the agricultural business, especially with respect to its connection infrastructure. The combination of smart devices, remote data collecting, the utilization of sensors and vehicles via the use of mobile devices and the internet, cloud-based intelligent analysis and decision-making, and the automation of agricultural activities have all been vital components of this phenomenon. The agricultural industry has undergone a great change as a result of the skills that have been developed in areas such as resource optimization, climate change effect mitigation, and crop yield improvement.

For the purpose of collecting and sending agricultural data throughout different development phases, researchers have suggested several methods, strategies, and instruments. These approaches, techniques, and tools vary according to the kind of crop and area. For the purpose of data collection and dissemination, a number of different firms provide communication tools, sensors, robots, huge machinery, and drones. Food and agricultural agencies, in addition to other governmental organizations, are working together to set standards and procedures for the use of technology in the preservation of the quality of food and the environment. There are several methods, including photoelectricity, electromagnetics, conductivity, and ultrasound, that are used in the process of analyzing the texture and structure of the soil, as well as its nutrient content, vegetation, humidity, vapor, air, temperature, and other elements linked to this topic. According to the findings of earlier studies, data collected by remote sensing may be employed to recognize crop varieties, categorize different kinds of pests and weeds, ascertain levels of stress in soil and plant conditions, and monitor levels of dryness.

The general health of a plant may be affected by a number of elements, including the amount of moisture in the soil, the availability of nutrients, the intensity of the light, the humidity, the amount of precipitation that falls, and the color of the leaves. The plants are being watched while the temperature and light conditions are kept at their ideal levels, and just a small amount of microirrigation is being applied. A wide variety of sensors are employed in order to detect a myriad of properties. According to the conclusions of the investigation, it has been noticed that the sensor is able to detect changes that are greater than a predefined threshold. This is something that has been seen. When the predetermined limit is breached, the sensor starts communicating the changes it has noticed with the microcontroller. The microcontroller will subsequently conduct the necessary actions in accordance with the predetermined protocol until the parameter reaches the level that is considered to be ideal [1].

A wide variety of sensors, such as those for temperature, humidity, soil patterns, airflow, location, carbon dioxide, pressure, light, and wetness, are routinely used in sensing systems. According to findings from earlier studies, the appropriateness of sensors for use in agricultural settings is dependent on a number of important features, including but not limited to dependability, storage capacity, mobility, robustness, breadth, and computational effectiveness. The usage of wireless sensors has developed into an essential component in the gathering of data pertaining to crops and the presentation of supplementary information. The integration of different kinds of sensors with more complex agricultural gear and equipment may be adapted to meet the particular demands of a given application in a manner that is completely customizable.

The exploitation of satellite images has the ability to unearth differences in crop yield and soil quality that are influenced by the topographical elements that are located in the surrounding area. In order to obtain precise management of the production ingredients that boost productivity and effectiveness, such as seeds, fertilizers, and pesticides, this goal must first be accomplished. One of the oldest methods used in agriculture is the practice of producing crops in a controlled environment, also known as smart farming. Some practices were seen to become more prevalent in nations that had more severe weather conditions during the course of the 20th century. When crops are grown inside, their vulnerability to the effects of the surrounding environment is significantly decreased. The growing of crops, which was formerly confined to locations with optimum circumstances, has become possible at any time and any location thanks to the usage of sensors and communication technologies. The production of crops in controlled settings is based on a variety of elements including the design of sheds and materials used to limit wind impacts, aeration systems, reliable monitoring data, decision-support systems, and other variables connected to agricultural production. In greenhouse settings, accurate monitoring of the environment's many different factors is a substantial problem. As a consequence of this, a number of different measurement points are needed in order to make accurate predictions about the numerous factors required for the regulation and maintenance of the ambient temperature. An IoT greenhouse makes use of sensors to monitor and measure inside characteristics such as temperature, light pressure, and humidity.

### Research Hypotheses

There is no significant difference between the application of IoT in smart farming and better crop management.

There is no significant difference between the application of IoT in smart farming and site-specific nutrition management.

There is no significant difference between the application of IoT in smart farming and Support in achieving strategic goals.

## ANALYSIS AND DISCUSSION

This part of the study involves in performing detailed data analysis based on the data collected by the researchers. The IBM SPSS package is used to make the analysis.

In Table 5.1, it is noted that 60.2% of the respondents were male, 35.2% were in the age group 31–40 years, 65.6% were living in a semi-urban area, 61.7% have 1–2 dependents to take care of, 50.8% were farmers, and 27.3% possess experience of less than 4 years.

### Discriminant Analysis

This part of the analysis explores understanding the significant differences between the independent and dependent variables, for this purpose discriminant analysis is used. The main objective of using discriminant analysis is to classify the various mutually exclusive data groups into one based on the predictor value.

**TABLE 5.1**
**Frequency Analysis**

| Gender Composition | Frequency | Percent |
|---|---|---|
| Male | 77 | 60.2 |
| Female | 51 | 39.8 |
| **Age Classification** | | |
| Less than 30 years | 37 | 28.9 |
| 31–40 years | 45 | 35.2 |
| 41–50 years | 15 | 11.7 |
| Above 50 years | 31 | 24.2 |
| **Place of Residence** | | |
| Semi-urban | 84 | 65.6 |
| Rural | 44 | 34.4 |
| **Number of Dependents** | | |
| 1–2 | 79 | 61.7 |
| 3 and above | 49 | 38.3 |
| **Current Role** | | |
| Farmer | 65 | 50.8 |
| Working in agricultural company | 45 | 35.2 |
| Land owner | 18 | 14.1 |
| **Work Experience** | | |
| Less than 4 years | 35 | 27.3 |
| 4–8 years | 30 | 23.4 |
| 8–12 years | 22 | 17.2 |
| 12–16 years | 9 | 7 |
| Above 16 years | 32 | 25 |

From Table 5.2 it is noted that the mean value of all the independent variables for agree and strongly agree is more than 4 and their corresponding standard deviation is less. This shows that most of the respondents agree with the statement that the application of IoT enables better crop management, site-specific nutrition management of crops, and support in achieving the strategic development goals (SDGs)

Based on the analysis, it is noted that the significance value of Wilks's lambda is 0.00 (Table 5.3). This shows that the independent variables possess significant difference to the dependent variable.

The analysis shows that the classification functions are high for the responses agree and strongly agree for all the independent variables considered in the study (Table 5.4).

From Figure 5.1, it is noted that the model has created a grouping commonly for the responses agree and strongly agree as they tend to provide similar responses about the application of IoT in smart farming.

**TABLE 5.2**
**Group Statistics**

| IoT in Sustainable Farming | Dependent Variables | Mean | Std. Dev. |
|---|---|---|---|
| Strongly Disagree | Better crop management | 2.00 | 0.00 |
| | Site-specific nutrition management | 2.00 | 0.00 |
| | Support in achieving strategic goals | 1.00 | 0.00 |
| Disagree | Better crop management | 1.75 | 0.45 |
| | Site-specific nutrition management | 1.50 | 0.52 |
| | Support in achieving strategic goals | 1.75 | 0.45 |
| Neutral | Better crop management | 3.00 | 0.00 |
| | Site-specific nutrition management | 3.19 | 0.40 |
| | Support in achieving strategic goals | 3.19 | 0.40 |
| Agree | Better crop management | 4.74 | 0.44 |
| | Site-specific nutrition management | 4.91 | 0.28 |
| | Support in achieving strategic goals | 4.49 | 0.66 |
| Strongly Agree | Better crop management | 4.77 | 0.42 |
| | Site-specific nutrition management | 4.89 | 0.32 |
| | Support in achieving strategic goals | 4.56 | 0.50 |
| Total | Better crop management | 4.20 | 1.12 |
| | Site-specific nutrition management | 4.30 | 1.19 |
| | Support in achieving strategic goals | 4.02 | 1.13 |

**TABLE 5.3**
**Test of Wilks's Lambda**

| Tests of Equality of Group Means | Wilks's Lambda | F | dF1 | dF2 | Sig. |
|---|---|---|---|---|---|
| Better crop management | 0.12 | 218.24 | 4.00 | 123.00 | 0.00 |
| Site-specific nutrition management | 0.08 | 351.15 | 4.00 | 123.00 | 0.00 |
| Support in achieving strategic goals | 0.22 | 111.98 | 4.00 | 123.00 | 0.00 |

**TABLE 5.4**
**Classification Function Coefficients**

| IoT in Sustainable Farming | Strongly Disagree | Disagree | Neutral | Agree | Strongly Agree |
|---|---|---|---|---|---|
| Better crop management | 13.02 | 10.85 | 18.79 | 29.92 | 30.07 |
| Site-specific nutrition management | 17.64 | 12.72 | 27.04 | 42.00 | 41.73 |
| Support in achieving strategic goals | 1.41 | 4.56 | 8.11 | 10.89 | 11.18 |
| (Constant) | –32.97 | –24.63 | –85.81 | –200.20 | –200.87 |

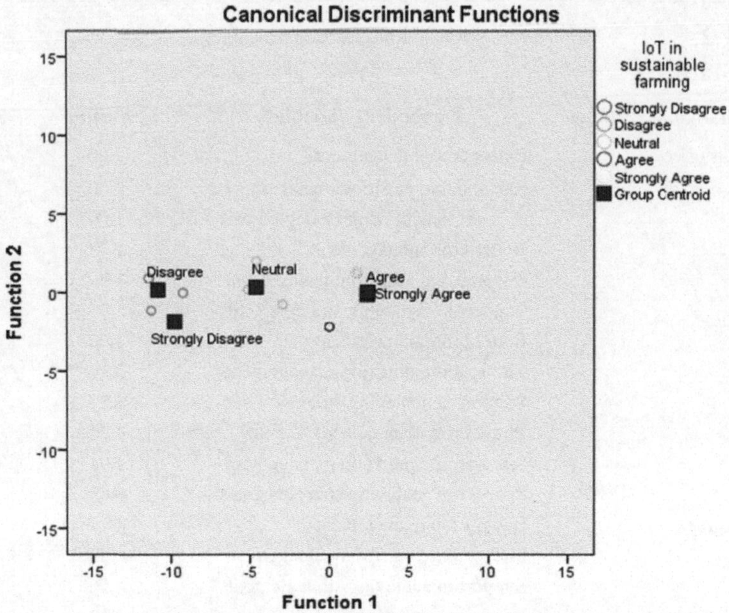


**FIGURE 5.1** Canonical discriminant grouping.

## TABLE 5.5
## Classification Analysis

| IoT in Sustainable Farming | Predicted Group Membership | | | | | |
|---|---|---|---|---|---|---|
| **Original** | **Strongly Disagree** | **Disagree** | **Neutral** | **Agree** | **Strongly Disagree** | **Total** |
| Strongly Disagree | 3 | 0 | 0 | 0 | 0 | 3 |
| Disagree | 3 | 9 | 0 | 0 | 0 | 12 |
| Neutral | 0 | 0 | 16 | 0 | 0 | 16 |
| Agree | 0 | 0 | 0 | 24 | 11 | 35 |
| Strongly Disagree | 0 | 0 | 0 | 30 | 32 | 62 |

65.6% of original grouped cases correctly classified

The analysis shows the comparison between the original values and the predicted values based on the model, it is noted that nearly 65.6% of the original cases are correctly predicted and classified and hence the model is stated to be a good fit (Table 5.5).

## CONCLUSION

In order to meet the issue of rapid global population growth on a shrinking amount of arable land, there is an urgent need for agricultural production systems that are both technologically sophisticated and highly efficient. It is of the utmost importance that people be aware of the significance of sustainable agriculture in ensuring that there will always be food available. This research aims to evaluate the elements that impact the view of farming as a respectable career choice among young people who are inventive and to investigate the potential of new technologies to improve agricultural output. Additionally, the study will investigate the factors that influence the perception of farming as a career choice among older generations. The focus of this research is on the investigation of various agricultural technologies, with a specific concentration on IoT, as a method of boosting agricultural productivity and intelligence in order to satisfy future needs. Academics and engineers might profit from the highlighting of the identification of probable future prospects and industrial constraints. Increasing agricultural productivity requires careful management of every square inch of field utilizing environmentally friendly sensors and communication technologies that are based on IoT.

The concepts of data analytics and machine learning are put to use in order to perform analyses on real-time data. The implementation of methods that are related to machine learning may be used to perform one of the most important aspects of crop production, which is the identification of genes that provide the best results. The use of machine learning in agriculture entails the prediction of genes that are most suited for agricultural output. This may be accomplished via the use of computer programs. This is especially important to keep in mind when picking seeds that are well suited to the unique environmental conditions of a given location. The use of algorithms for machine learning has the ability to recognize things that are in great demand as well as those that are presently out of supply. Recent developments in machine learning and analytics have made it possible for farmers to properly classify their crops before processing them and distributing them to customers.

## BIBLIOGRAPHY

1. Adamides, G.; Kalatzis, N.; Stylianou, A.; Marianos, N.; Chatzipapadopoulos, F.; Giannakopoulou, M.; Papadavid, G.; Vassiliou, V.; Neocleous, D. Smart Farming Techniques for Climate Change Adaptation in Cyprus. *Atmosphere* 2020, 11, 557.
2. Ayaz, M.; Ammad-uddin, M.; Baig, I.; Aggoune, E.M. Wireless Sensor's Civil Applications, Prototypes, and Future Integration Possibilities: A Review. *IEEE Sens. J.* 2018, 18, 4–30.
3. Ferrández-Pastor, F.J.; García-Chamizo, J.M.; Nieto-Hidalgo, M.; Mora-Martínez J. Precision Agriculture Design Method Using a Distributed Computing Architecture on Internet of Things Context. *Sensors* 2018, 18, 1731.
4. Hassan, A.; Aitazaz, A.F.; Farhat, A.; Bishnu, A.; Travis, E. Precision Irrigation Strategies for Sustainable Water Budgeting of Potato Crop in Prince Edward Island. *Sustainability* 2020, 12, 2419.

5. Hernández-Ochoa, I.M.; Gaiser, T.; Kersebaum, K.C.; Webber, H.; Seidel, S.J.; Grahmann, K.; Ewert, F. Model-Based Design of Crop Diversification Through New Field Arrangements in Spatially Heterogeneous Landscapes. A review. *Agron. Sustain. Dev.* 2022, 42, 74
6. Liakos, K.G.; Busato, P.; Moshou, D.; Pearson, S.; Bochtis, D. Machine Learning in Agriculture: A Review. *Sensors* 2018, 18, 2674.
7. Muhammad, S.F.; Shamyla, R.; Adnan, A.; Tariq, U.; Yousaf, B.Z. Role of IoT Technology in Agriculture: A Systematic Literature Review. *Electronics* 2020, 9, 319.
8. Potamitis, I.; Rigakis, I.; Tatlas, N.A.; Potirakis, S. In-Vivo Vibroacoustic Surveillance of Trees in the Context of the IoT. *Sensors* 2019, 19, 1366.
9. Quy, V.K.; Hau, N.V.; Anh, D.V.; Quy, N.M.; Ban, N.T.; Lanza, S.; Randazzo, G.; Muzirafuti, A. IoT-Enabled Smart Agriculture: Architecture, Applications, and Challenges. *Appl. Sci.* 2022, 12, 3396.
10. Reddy, T.; Dutta, M. Impact of Agricultural Inputs on Agricultural GDP in Indian Economy. *Theor. Econ. Lett.* 2018, 8, 1840–1853.
11. Shi, X.; An, X.; Zhao, Q.; Liu, H.; Xia, L.; Sun, X.; Guo, Y. State- of- the- Art Internet of Things in Protected Agriculture. *Sensors* 2019, 19, 1833.
12. Zhmud, V.A.; Kondratiev, N.O.; Kuznetsov, K.A.; Trubin, V.G.; Dimitrov, L.V. Application of Ultrasonic Sensor for Measuring Distances in Robotics. *J. Phys. Conf. Ser.* 2018, 1015, 032189.

# 6 Blockchain-Based Parking System Using IoT

*S Sahunthala, Latha Parthiban, and R Parthiban*

## INTRODUCTION

The various components of a smart city are shown in Figure 6.1. These components are important for creating a smart city, and the Internet of Things (IoT) is an effective way to handle this.

The IoT evolution is shown in Figure 6.2. It has helped in the machine-to-machine interface. The main motivation behind the use of IoT is the integration of devices.

The workings of IoT are presented in Figure 6.3.

## LITERATURE SURVEY

Parking has become an important requirement today. Some parking systems have tried using RFID for managing both single and multistoried parking lots [1, 2]. Researchers have proposed a smart parking system that is secure and automated [3–8]. An efficient automated parking system using GSM and Zigbee technology has also been studied [9, 10].

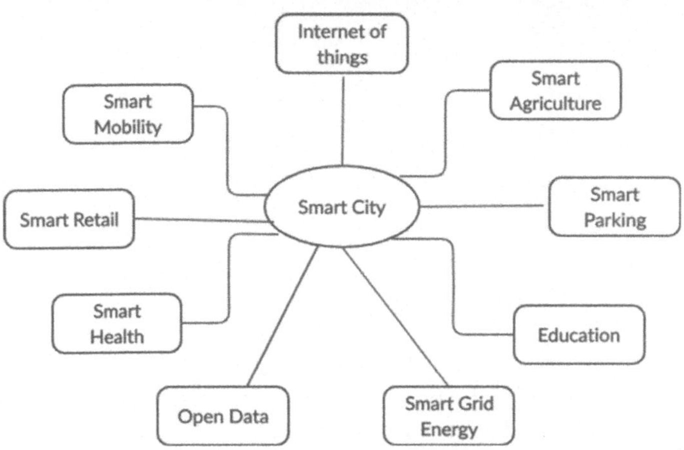


**FIGURE 6.1** Example of smart city components.

DOI: 10.1201/9781003428886-8

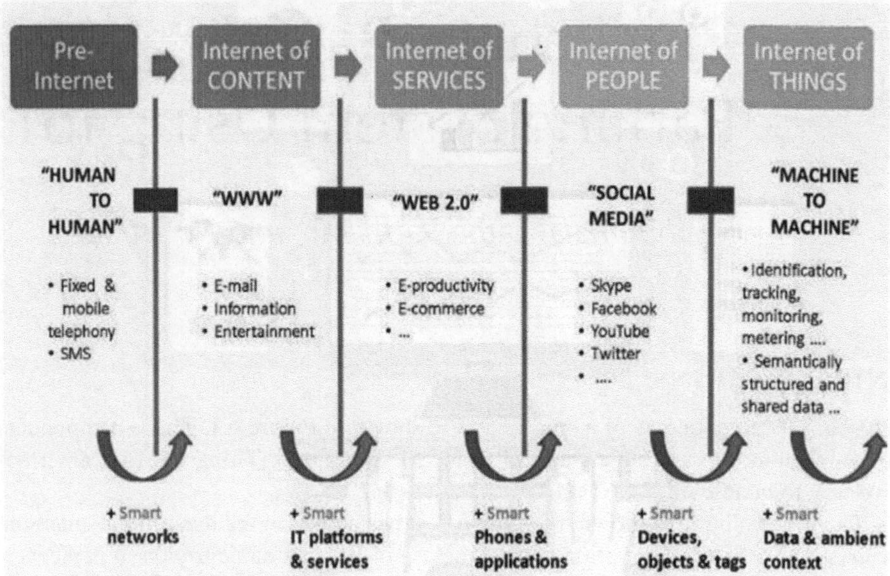


FIGURE 6.2 IoT evolution.

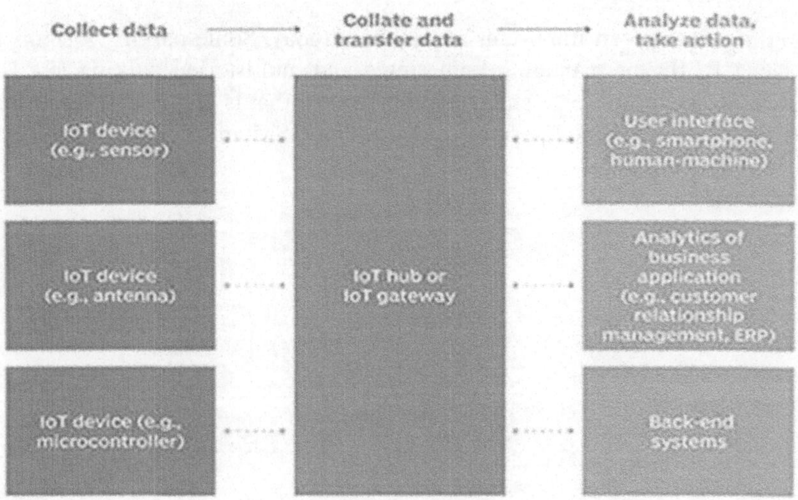


FIGURE 6.3 How IoT works.

WSN-based secure parking [11–13] as well as an agent-based parking system platform have been suggested [14]. The Google Maps application has been used to search for parking area [15, 16]. Ultrasonic sensors are used to sense the parking slots. If the LED display is on, then there is a free parking slot; if the display is off, then the slot is filled.

## SYSTEM ANALYSIS

This research work helps to facilitate users and admins to park vehicles in a well-defined method. This study helps to show updates about empty parking slots in the location chosen by the users. The research uses Arduino to monitor and control the parking system. The Wi-Fi module ESP8266 NodeMCU is used to communicate between the IoT and the web server. The web server updates the user about empty parking slots every three minutes.

Infrared (IR) sensors are affixed to the entry gate and exit gates to sense objects and when to open the gate. The servomotors collect the signal from the IR sensors in the entry and exit gates. The IR sensors used in the parking slots sense the status of the slots and send a signal to Arduino. The LCD display at the entry gate will display the number of empty and filled slots in the parking area. Users can check slot status through the web server. Users can choose to display the status of their desired parking location.

## REQUIREMENT SPECIFICATION

### Hardware Requirements

The various hardware requirements are shown in Figures 6.4 to 6.8.

### Software Requirements

- Tools required: Arduino IDE (Figure 6.9)

Operating system: Linux OSPre requisite JDK 1.6 or above

**FIGURE 6.4** Arduino Uno.

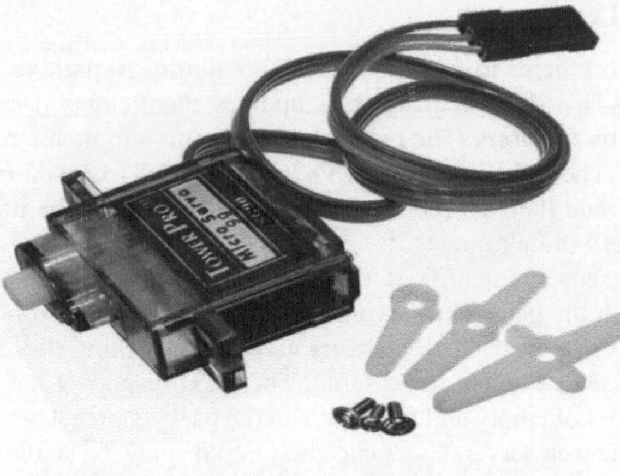

**FIGURE 6.5** Servomotor.

**FIGURE 6.6** LCD display.

**FIGURE 6.7** Infrared sensor.

## SYSTEM DESIGN

### Design Overview

The characters involved are user and admin. The situation is to choose the location and parking area to get updates about parking slot status (Figure 6.10).

FIGURE 6.8 ESP8266 NodeMCU.

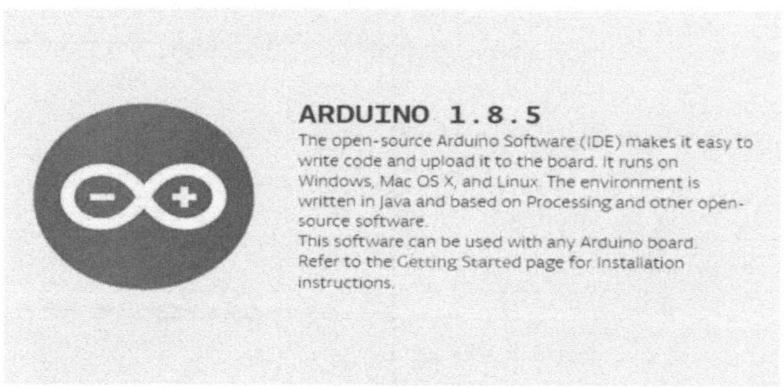


FIGURE 6.9 Arduino IDE.

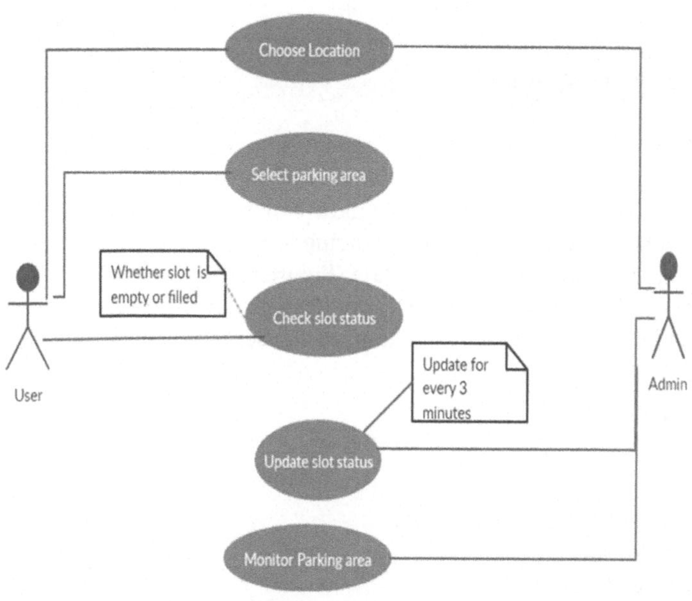


FIGURE 6.10 Use case diagram.

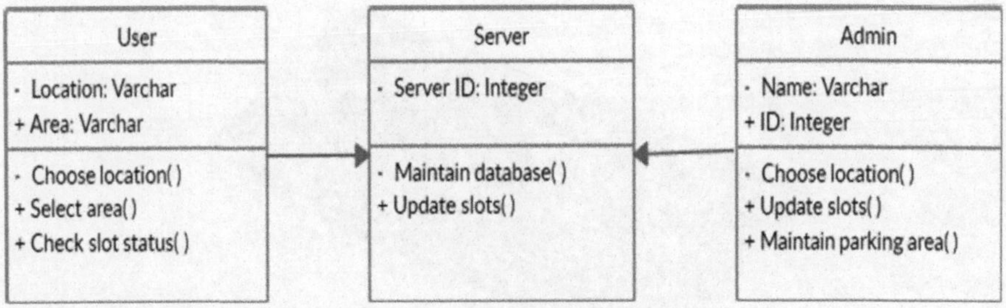


**FIGURE 6.11** Class diagram.

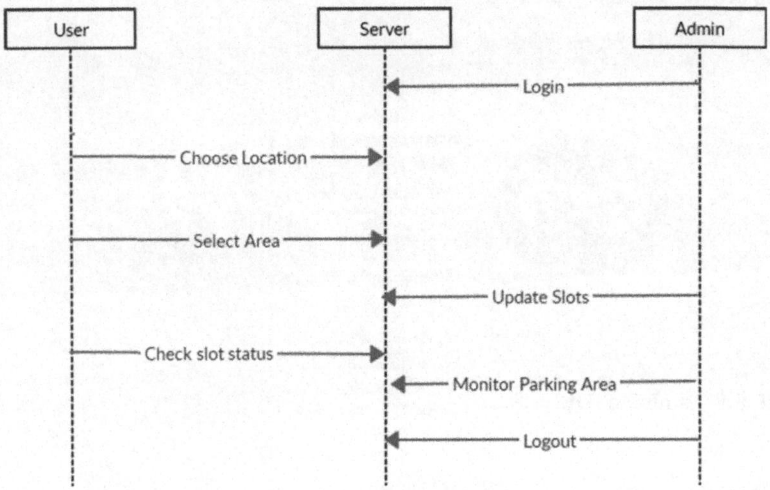


**FIGURE 6.12** Sequence diagram.

Description: The classes and interfaces in the diagram represent the attributes and operations. The classes involved are user, server, and admin (Figure 6.11).

Description: The admin provides the information in both modules. The hardware interface displays information about the number of empty and filled slots and it will also be updated in the web server module (Figure 6.12).

Description: The step-by-step process of the parking system is displayed in Figure 6.13. The activity flow includes both admin and user processes.

## IMPLEMENTATION

### Modules

- Selection of location and area.
- Check slot status.
- Updating of lots.
- Visualization.

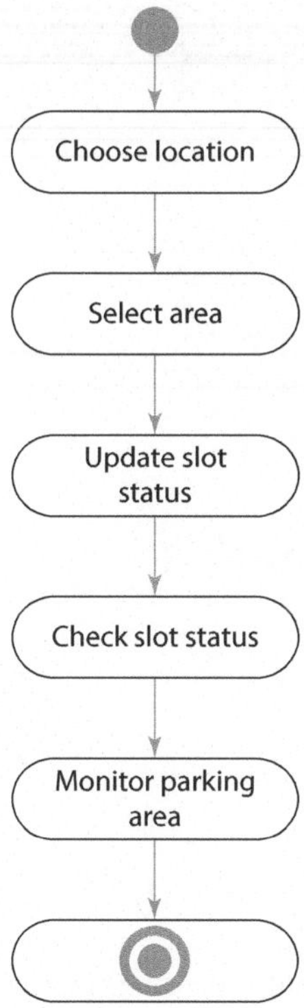


**FIGURE 6.13** Activity diagram.

## Architecture

The architecture represents the overall working of the parking system (Figure 6.14). It is implemented in two modules: hardware implementation and web server. Both modules update the status of the parking slots and display the output on a screen. In the first module, we implement the hardware component to develop the parking system using the Arduino board. The Arduino board works as a microcontroller and it controls the overall system. There is an entrance and exit gate in the parking system that is used to allow the vehicle to enter the system.

The IR sensors are used to sense the vehicle and to send a signal to the Arduino about the vehicle's status. The Wi-Fi module is used to integrate the Arduino with the web server that updates the parking slot status.

The web server is developed using Arduino IDE, which is used for updating the parking slot status for users. In the web server, users can select the location and particular parking area. When the particular area is chosen, the web server will display the slot status to the users.

The slot details are stored in the database to maintain the records. The admin can monitor the parking system regularly and update the status every three minutes. This system will help the user to know the current availability status and reduce the time to search for parking.

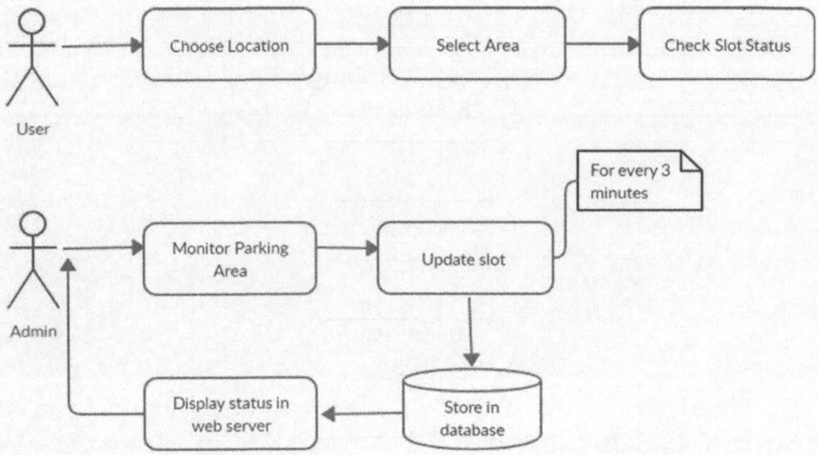


**FIGURE 6.14** Architecture diagram.

## Modules Description

### Selection of Location and Area

This module is used to select the location and area to search for parking slots. The user will choose the location. This module will be accessed by using GPS to select the location. By using this module, the user can check the status in the area they desire.

### Check Slot Status

The slot status is displayed on the LCD and on the web server. The LCD display will be affixed in the parking system and used to display the status to users who are entering the parking system. The web server is for mobile users and displays the status which integrates with the parking system.

### Updating of Slots

This module can be controlled by the admin side. The admin will monitor the parking area and update the parking slot status every three minutes. The slot details will be stored in the database server.

### Visualization

The generated results are visualized in a web server, which makes the parking process easier. The web server is monitored and controlled using the Arduino.

# TESTING

## Manual Testing

Table 6.1 provides the manual testing of the project and Table 6.2 its implementation. Table 6.3 provides the testing of hardware requirements.

**TABLE 6.1**
**Manual Testing**

| Test Case ID | Test Description | Test Procedure | Test Input | Expected Result | Actual Result |
|---|---|---|---|---|---|
| T1-1 | To check whether Arduino IDE is installed in the system. | Open command prompt in the system. | Type Arduino IDE version. | The Arduino version must be displayed. | The Arduino version is displayed. |
| T1-2 | To check whether Arduino IDE opens correctly. | Open the Arduino IDE. | Start typing the Arduino code. | The editor must be opened and the coding must be typed. | The editor was opened and the code was correctly typed. |
| T1-3 | To check whether it implemented the code for Arduino board. | Compile the Arduino code written. | Enter command for compile. | The code must be compiled. | The code was compiled. |
| T1-4 | To check whether the Arduino code is executed. | Execute the Arduino code written. | Enter the command to execute. | The code must be executed. | The code was executed. |
| T1-5 | To check whether the code is implemented on the web server. | Execute the Arduino code written. | Enter the command for compile. | The code must be compiled. | The code was compiled. |
| T1-6 | To check whether the implemented web server code is executed. | Execute the web server code written. | Enter the command to execute. | The code must be executed. | The code was executed. |
| T1-7 | To check whether the web server integrates with the Arduino board. | Compile the web server code. | Compile the web server code. | The web server displayed the status in the Arduino board correctly. | The web server displayed the status in the Arduino board correctly. |
| T1-8 | To check whether the web server displays the slot status. | Compile the web server code. | Compile the web server code. | The slot status must be displayed correctly. | The slot status was displayed correctly. |

*(Continued)*

**TABLE 6.1 (CONTINUED)**
**Manual Testing**

| Test Case ID | Test Description | Test Procedure | Test Input | Expected Result | Actual Result |
|---|---|---|---|---|---|
| T1-9 | To check whether the server allows user to choose the location. | The user must choose the location. | The user must choose location from the drop-down box only. | The user can to choose the location. | The user chose the location. |
| T1-10 | To check whether the server allows user to select the particular area. | The user must select the particular area. | The user must select area from the drop-down box only. | The user can to select the area. | The user selected the particular area. |
| T1-11 | To check whether the system updates every 3 minutes. | Execute the web server code. | The admin must update the slot every 3 minutes. | The server updates every 3 minutes. | The server updated every 3 minutes. |

**TABLE 6.2**
**Testing for Arduino Implementation**

| Test Case ID | Test Description | Test Procedure | Test Input | Expected Result | Actual Result |
|---|---|---|---|---|---|
| T2-1 | To check whether the Arduino board works properly. | Give power supply and test the board. | Give input connections for Arduino board. | The Arduino must be connected and work properly. | The Arduino was connected and works properly. |
| T2-2 | To check whether the ESP8266 NodeMCU works properly. | Give power supply and test the ESP8266 NodeMCU. | Give input connections for ESP8266 NodeMCU. | The ESP8266 NodeMCU must be connected and works properly. | The ESP8266 NodeMCU was connected and works properly. |
| T2-3 | To check whether the servomotor works properly. | Give power supply and test the servomotor. | Give input connections for servomotor. | The servomotor must be connected and works properly. | The servomotor was connected and works properly. |
| T2-4 | To check whether the infrared (IR) sensors work properly. | Give power supply and test infrared (IR) sensors. | Give input connections for infrared (IR) sensors. | The infrared (IR) sensors must be connected and work properly. | The infrared (IR) sensors were connected and work properly. |
| T2-5 | To check whether the LCD display works properly. | Give power supply and test the LCD display. | Give input connections for the LCD display. | The LCD display must be connected and work properly. | The LCD display was connected and work properly. |
| T2-6 | To check whether the servomotor opens and closes an entrance gate. | Sensor will sense the object in the entrance gate. | An object must enter the entrance gate. | The entrance gate must open and close. | The entrance gate opens and closes. |
| T2-7 | To check whether the servomotor opens and closes an exit gate. | Sensor will sense the object in the exit gate. | An object must enter the exit gate. | The exit gate must open and close. | The exit gate opens and closes. |

*(Continued)*

**TABLE 6.2 (CONTINUED)**
**Testing for Arduino Implementation**

| Test Case ID | Test Description | Test Procedure | Test Input | Expected Result | Actual Result |
|---|---|---|---|---|---|
| T2-8 | To check whether the IR sensors sense the object correctly. | The object is placed near the sensor. | The sensor will sense the object. | The sensor must sense object and sends signal to Arduino board. | The sensor must sense object and sends signal to Arduino board. |
| T2-9 | To check whether the LCD displays the slot status correctly. | Give power supply and test the input. | The object must be placed in the slot to check the LCD. | The LCD must display the slot status correctly. | The LCD displayed the slot status correctly. |
| T2-10 | To check whether the ESP8266 NodeMCU integrates with the web server. | Compile the web server code. | Compile the web server code. | The web server display the status in the Arduino board correctly. | The web server displayed the status in the Arduino board correctly. |

**TABLE 6.3**
**Testing for Hardware Requirements**

| Test Case ID | Description | Status |
|---|---|---|
| T1-1 | To check whether Arduino IDE is installed in the system. | PASS |
| T1-2 | To check whether Arduino IDE opens correctly. | PASS |
| T1-3 | To check whether it implemented the code for Arduino board. | PASS |
| T1-4 | To check whether the implemented Arduino board code is executed. | PASS |
| T1-5 | To check whether code is implemented on the web server. | PASS |
| T1-6 | To check whether the implemented web server code is executed. | PASS |
| T1-7 | To check whether the web server integrates with the Arduino board. | PASS |
| T1-8 | To check whether the web server displays the slot status. | PASS |
| T1-9 | To check whether the server allows user to choose the location. | PASS |
| T1-10 | To check whether the server allows the user to select a particular area. | PASS |
| T1-11 | To check whether the system updates every 3 minutes. | PASS |
| T2-1 | To check whether the Arduino board works properly. | PASS |
| T2-2 | To check whether the ESP8266 NodeMCU is working properly. | PASS |
| T2-3 | To check whether the servomotor is working properly. | PASS |
| T2-4 | To check whether the infrared (IR) sensors work properly. | PASS |
| T2-5 | To check whether the LCD display is working properly. | PASS |
| T2-6 | To check whether the servomotor opens and closes an entrance gate. | PASS |
| T2-7 | To check whether servomotor opens and closes an exit gate . | PASS |
| T2-8 | To check whether IR sensors sense the object correctly. | PASS |
| T2-9 | To check whether LCD displays the slot status correctly. | PASS |
| T2-10 | To check whether ESP8266 NodeMCU integrates with the web server. | PASS |

## SOURCE CODE

### Arduino Implementation

```
#include <Servo.h> #include <LiquidCrystal.h>
LiquidCrystal lcd(12, 11, 5, 4, 3, 2); const int analogInPin
= A0;
int sensorValue = 0; int servoPin = 10; int i = 90;
Servo servo;
const int analogInPin1 = A1; int sensorValue1 = 0;
int servoPin1 = 9; int i1 = 90;
Servo servo1;
const int analogInPin2 = A2; int sensorValue2 = 0;
const int analogInPin3 = A3; int sensorValue3 = 0;
const int analogInPin4 = A4;
int sensorValue4 = 0;
const int analogInPin5 = A5; int sensorValue5 = 0;
int LIGHT1 = 13; int LIGHT2 = 8; int LIGHT3 = 7; int LIGHT4
= 6;
```

```
void setup()
{
Serial.begin(9600); pinMode(LIGHT1, OUTPUT); pinMode(LIGHT2,
OUTPUT); pinMode(LIGHT3, OUTPUT); pinMode(LIGHT4, OUTPUT);
servo.attach(servoPin); servo1.attach(servoPin1); lcd.beg
in(16, 2);
}
void loop()
{
lcd.setCursor(1,1); lcd.println("FILL ");
lcd.setCursor(0,0); lcd.println("EMPTY ");
sensorValue2 = analogRead(analogInPin2); if(sensorValue2 <
600){
digitalWrite(LIGHT1, HIGH); Serial.println(LIGHT1); Serial.
println(sensorValue2); lcd.setCursor(6,1); lcd.println("1 ");
}
else{
digitalWrite(LIGHT1, LOW); Serial.println(LIGHT1); Serial.
println(sensorValue2); lcd.setCursor(6,0); lcd.println("1 ");
}
sensorValue3 = analogRead(analogInPin3); if(sensorValue3 <
600){
digitalWrite(LIGHT2, HIGH); lcd.setCursor(8,1);
lcd.println("2 ");
}
else{
digitalWrite(LIGHT2, LOW); lcd.setCursor(8,0); lcd.println("2
");
}
sensorValue4 = analogRead(analogInPin4); if(sensorValue4 <
600){
digitalWrite(LIGHT3, HIGH); lcd.setCursor(10,1); lcd.
println("3 ");
}
else{
digitalWrite(LIGHT3, LOW); lcd.setCursor(10,0); lcd.println("3
");
}
sensorValue5 = analogRead(analogInPin5); if(sensorValue5 <
600){
digitalWrite(LIGHT4, HIGH); lcd.setCursor(12,1);
lcd.println("4 ");
}
else{
digitalWrite(LIGHT4, LOW); lcd.setCursor(12,0); lcd.println("4
");
}
```

```
sensorValue = analogRead(analogInPin); if(sensorValue1<700){
if (int i =0; i<=90){ servo.write(i); delay(15);
}
}
else{
if (int i = 90; i>= 0) { servo.write(i); delay(15);
}
}
sensorValue1 = analogRead(analogInPin1); if(sensorValue<700){
if (int i1 =0; i1<=90){ servo1.write(i1); delay(15);
}
}
else{
if (int i1 = 90; i1>= 0) { servo1.write(i1); delay(15);
}
}
}
```

## Web Server Implementation

```
#include <ESP8266WiFi.h> #include <WiFiClient.h>
#include <ESP8266WebServer.h> const int ProxSensor=16;
char values;
const int ProxSensor1=5; char values1;
const int ProxSensor2=4; char values2;
const int ProxSensor3=0; char values3;
const char* ssid = "kousigan";
double data;
String readcurrent()
{
long state = digitalRead(ProxSensor); if(state == HIGH)
{
Serial.println(state); values='E';
}
else
{
Serial.println(state);
values='F';
}
return String(values);
}
String readcurrent1()
{
long state1 = digitalRead(ProxSensor1); if(state1 == HIGH)
{
Serial.println(state1); values1='E';
}
```

```
else
{
Serial.println(state1); values1='F';
}
return String(values1);
}
String readcurrent2()
{
long state2 = digitalRead(ProxSensor2); if(state2 == HIGH)
{
Serial.println(state2); values2='E';
}
else
{
Serial.println(state2); values2='F';
}
return String(values2);
}
String readcurrent3()
{
long state3 = digitalRead(ProxSensor3); if(state3 == HIGH)
{
Serial.println(state3);
values3='E';
}
else
{
Serial.println(state3); values3='F';
}
return String(values3);
}
<head>
<title>Font Awesome Icons</title>
<meta name="viewport" content="width=device-width,
initial- scale=1">
<link rel="stylesheet" href="https://cdnjs.cloudflare.com/ajax
/libs/font- awesome/4.7.0/css/font-awesome.min.css">
<style>
body {
background-image: url('https://c4.wallpaperflare.com/wallpaper
/590/532/911/colors-of- the-dawn-wallpaper-preview.jpg');
background-repeat: no-repeat; background-attachment: fixed;
background-size: cover;
}
</style>
</head>
<body>
```

```
<center><font size="11"><p style="color:red;">CAR PARKING</
p></font><center>
<a></a>
<center><p>
<i class="fas fa-car-side" style="font-
size:48px;color:black"> <font size="6"><span class="ds-
labels">CAR PARK1 =</span> <d><span style="color:blue;"
id="values"></span></d></font>
</p></center>
<center><p>
<i class="fas fa-car-side" style="font-
size:48px;color:black"> <font size="6"><span class="ds-
labels">CAR PARK2 =</span> <d><span style="color:blue;"
id="values1"></span></d></font>
</p></center>
<center><p>
<i class="fas fa-car-side" style="font-
size:48px;color:black"> <font size="6"><span class="ds-
labels">CAR PARK3 =</span> <d><span style="color:blue;"
id="values2"></span></d></font>
</p></center>
<center><p>
<i class="fas fa-car-side" style="font-
size:48px;color:black"></i> <font size="6"><span class="ds-
labels">CAR PARK4 =</span> <d><span style="color:blue;"
id="values3"></span></d></font>
</p></center>
</body>
};
xhttp.open("GET", "/values3", true); xhttp.send();
}, 1) ;
</script>
</html>)rawliteral";
String processor(const String& var)
{
if(var == "CAR PARK1")
{
return readcurrent();
}
if(var == "CAR PARK2")
{return readcurrent1();
}
if(var == "CAR PARK3")
{return readcurrent2();
}
if(var == "CAR PARK4")
{return readcurrent3();
```

```
}
return String();
}
void setup(){ pinMode(ProxSensor,INPUT); Serial.begin(115200);
WiFi.begin(ssid, password); //begin WiFi connection Serial.
println("");
// Wait for connection
while (WiFi.status() != WL_CONNECTED) { delay(500);
Serial.print(".");
}
Serial.println(""); Serial.print("Connected to "); Serial.
println(ssid); Serial.print("IP address: "); Serial.
println(WiFi.localIP()); server.on("/", HTTP_GET, []()
{
server.send(200, "text/html", index_html);
});
server.on("/values", HTTP_GET, []()
{
server.send(200, "text/plain", readcurrent());
});
server.on("/values1", HTTP_GET, []()
{
server.send(200, "text/plain", readcurrent1());
});
server.on("/values2", HTTP_GET, []()
{
server.send(200, "text/plain", readcurrent2());
});
server.on("/values3", HTTP_GET, []()
{
server.send(200, "text/plain", readcurrent3());
});
server.begin();
Serial.println("Web server started!");
}
void loop(){ server.handleClient();
```

## CONCLUSION AND FUTURE ENHANCEMENTS

This research work has implemented an IoT-based parking system that checks the status of a parking slot online and implemented it through a created web server for checking the status of the parking slots. Users can even add additional features to the server. As a future enhancement, the researchers would like to improve the efficiency of the server and add features like reserving parking slots, navigating the user from the entrance to the slot, and payment options. This will be helpful for the users to search for parking spaces and it will be user-friendly.

## REFERENCES


1. Mazlan, Muhammad & A Hamid, Isredza Rahmi and Kamaludin, Hazalila. 'Radio Frequency Identification (RFID) Based Car Parking System', JOIV: International Journal on Informatics Visualization, vol. 2, no. 4–2, 2018.
2. Aye, K. Nan, Zin Oo, P. and Naing, W. War. 'RFID Based Automatic Multi-storeyed Car Parking System', International Journal of Science and Engineering Applications, Vol. 8, no. 07, pp. 172–175, 2019.
3. Abdulkader, Omar, et al. 'A Novel and Secure Smart Parking Management System (SPMS) Based on the Integration of WSN, RFID, and IoT', 15th Learning and Technology Conference (L&T). IEEE, 2018.
4. Razinkoval, Anastasia, Chol, Hyun-Chan, and Jeon, Hong-Tae. 'An Intelligent Auto Parking System for Vehicle', International Journal of Fuzzy Logic and Intelligent Systems, vol. 12, no. 3, pp. 226–231, 2012.
5. Azshwanth, D., Koshy, Mithul Titten and Balachander, T. 'Automated Car Parking System', International Conference on Physics and Photonics Processes in Nano Sciences, vol. 1362, pp. 1–9, 2019.
6. Patil, M. and Bhonge, V. N. 'Wireless Sensor Network and RFID for Smart Parking System', International Journal of Emerging Technology and Advanced Engineering, vol. 3, no. 4, pp. 188–192, 2013.
7. DharmaReddy, P., RajeshwarRao, A. and Ahmed, Dr. Syed Musthak. 'An Intelligent Parking Guidance and Information System by Using Image Processing Technique', International Journal of Advanced Research in Computer and Communication Engineering, Vol. 2, no. 10, 2013.
8. Al-Kharusi, H. and Al-Bahadly, I. 'Intelligent Parking Based on Image Processing', World Journal of Engineering and Technology, vol. 2, no. 02, pp. 55–67, 2014.
9. Rahayu, Y. and Mustapa, F. N. 'A Secure Parking Reservation System using GSM Technology', International Journal of Computer and Communication Engineering, vol. 2, no. 4, p. 518, 2013.
10. Yee, H. C. and Rahayu, Y. 'Monitoring Parking Space Availability via ZigBee Technology', International Journal of Future Computer and Communication, vol. 3, no. 6, p. 377, 2014.
11. Poojaa, A. 'WSN-Based Secure Vehicle Parking Management and Reservation System,' National Conference on Research Advances in Communication, Computation, Electrical Science and Structures (NCRACCESS-2015), 2015.
12. Karbab, E., Djenouri, D., Boulkaboul, S. and Bagula, A. 'Car Park Management with Networked Wireless Sensors and Active RFID.' In 2015 IEEE International Conference on Electro/Information Technology (EIT), pp. 373–378. IEEE, 2015, May.
13. Orrie, O., Silva, B. and Hancke, G. P. 'A Wireless Smart Parking System', In Industrial Electronics Society, IECON 2015- 41st Annual Conference of the IEEE, pp. 004110–004114. IEEE, 2015, November.
14. Chou, S. Y., Lin, S. W. and Li, C. C. 'Dynamic Parking Negotiation and Guidance Using an Agent-Based Platform', Expert Systems with Applications, vol. 35, no. 3, pp. 805–817, 2008.
15. Basavaraju, S. R. 'Automatic Smart Parking System using Internet of Things (IoT)', International Journal of Scientific and Research Publications, vol. 5, no. 12, pp. 629–632, 2015.
16. Gandhi, B. M. K. and Rao, M. K. 'A Prototype for IoT based Car Parking Management System for Smart Cities', Indian Journal of Science and Technology, vol. 9, no. 17, pp. 1–6, 2016.

# 7 Blockchain-Based Online Attendance Management System

*Latha Parthiban, S Sahunthala, and R Parthiban*

## INTRODUCTION

The practice of marking student attendance is traditionally done using a logbook. New methods of marking attendance have evolved with automation. But special care must be taken to store the data safely from later, unauthorized modification. Many applications use the Internet of Things (IoT) to capture attendance with biometrics. Parents can access the attendance of their ward and track a problem if there is an issue. For scholarships, the government can directly access the data.

A blockchain is made up of a growing list of records called blocks that are linked by encryption. Many blockchain frameworks have been proposed to integrate it with industrial advancements. Again integrating it with federated learning helps it work collaboratively in a distributed environment.

## LITERATURE SURVEY

Avoade et al. [1] created a trusted blockchain environment for managing data. Androulaki et al. [2] created a Hyperledger fabric for blockchain. Aazam et al. [3] designed an internet cloud of things and resolved most problems in integration. Makhdoom et al. [4] proposed methods and solutions for the use of blockchain. Ramezan et al. [5] proposed a routing protocol for IoT, and Lim et al. [6] proposed an automated attendance system using IoT. Arbain et al. [7] presented an attendance system based on the web. Chintalapati et al. [8] developed a face recognition-based attendance system. A biometric attendance management system [9, 10] and blockchain-based attendance system [11–13] have also been developed and implemented [14, 15] .

By using an employee management system, one can create an identity profile for an employee. The profile can be managed (e.g., editing, creating, deleting) in the system. Mostly this type of system is highly secure. If the employee wants to transfer to another branch, management can help by sending an application for a branch transfer request. If the employee wants to apply for leave, they can request it through

 DOI: 10.1201/9781003428886-9

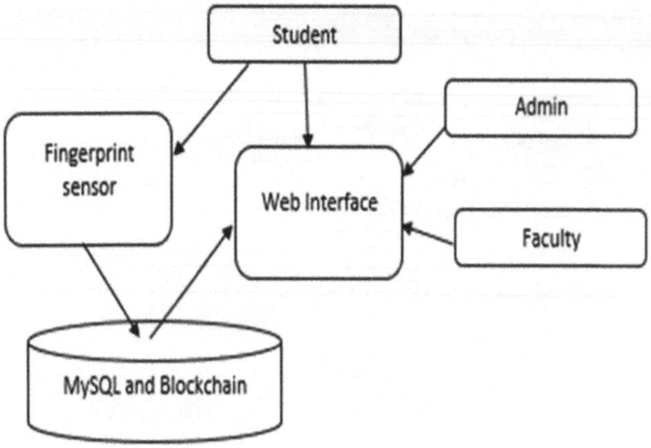


**FIGURE 7.1** System architecture.

the system. In a university environment, faculty become the employees and they can use all the facilities as shown in Figure 7.1.

## SYSTEM ANALYSIS

An online attendance management system was developed with PHP. This online management system has several modules, including signup, login, teacher, student, admin, and attendance. In the signup module, the user can register in the system by providing the required details. For example, the e-mail column can only be completed with a valid mail ID. The password should be eight characters long and a combination of uppercase and lowercase letters. The username and password must not be the same.

In the login module, the user can log in to the online attendance management system by giving their username and password. In the teacher module, the teacher can log in with their credentials to mark the daily attendance of students. In the student module, the student also can log in with credentials. But they can't edit the marked attendance. If there is a mistake in the marked attendance, students can rectify it by informing the teachers. In the admin module, the admin can also log in by giving their credentials. They can check attendance status by date, and CRUD (create, read, update, delete) operations of the student list are performed. The attendance report counts the presence and absence of students.

The proposed system helps to mark the student's attendance without any buffering. Unlike PHP, there is no flexibility when compared to MERN stack. So, the web application is created with the help of MERN stack (MongoDb, Expressjs, Nodejs, Reactjs).

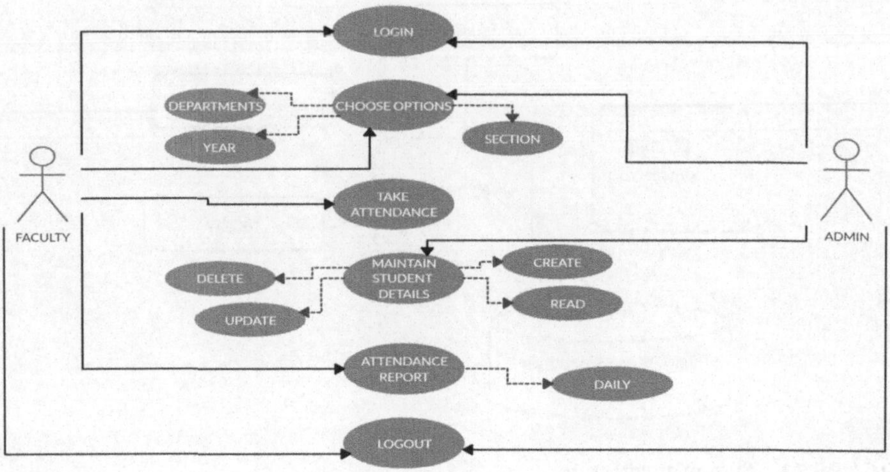


**FIGURE 7.2** Use case diagram.

## REQUIREMENTS SPECIFICATION

The use case diagram for the proposed system is provided in Figure 7.2. The characters involved here are faculty and admin. The setting is to gather the details from the students and attendance analysis. The analysis is done using the MERN stack. The class diagram is presented in Figure 7.3 and the sequence diagram in Figure 7.4.

The student provides the information in both modules. In the Weka analysis module, the data collected from the interface is put into the database, which in turn is imported to the Weka environment for analysis. The activity diagram is shown in Figure 7.5.

The data is obtained from the students and analyzed using the two modules, namely, the Java implementation and Weka analysis. The naive Bayes algorithm is employed for analysis and the results are visualized.

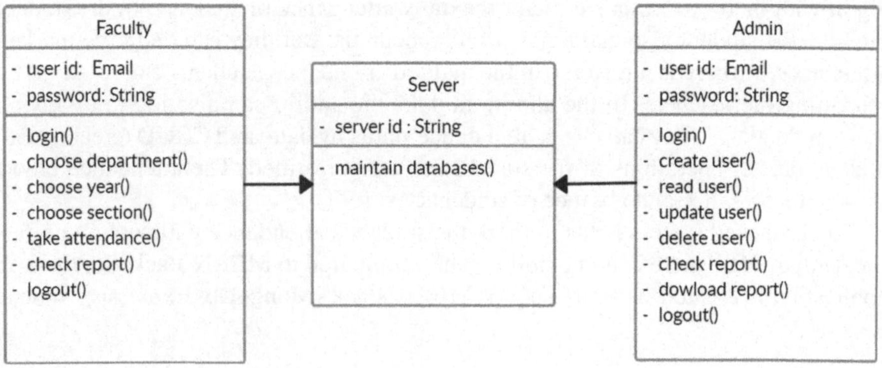


**FIGURE 7.3** Class diagram.

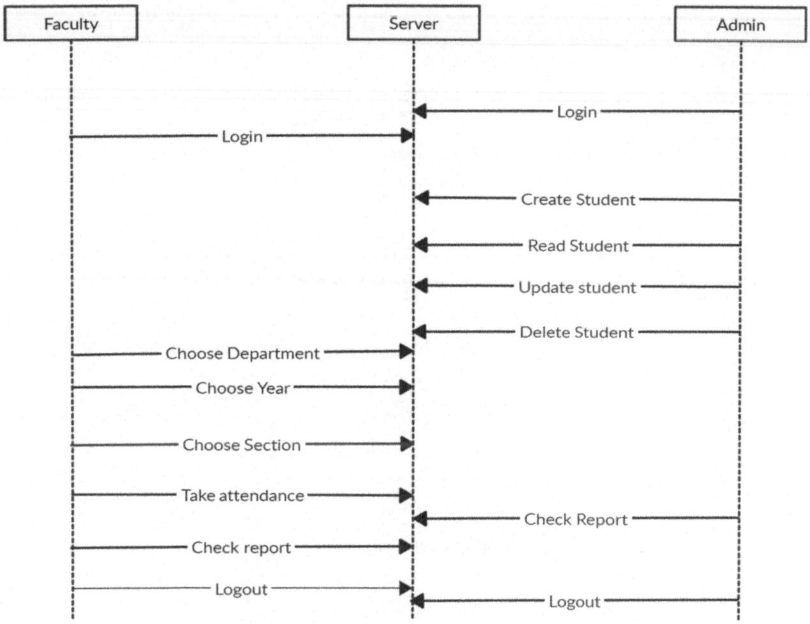


**FIGURE 7.4** Sequence diagram.

## SYSTEM IMPLEMENTATION

### Modules in System Implementation

Faculty module
Admin module

### Architecture Diagram

The architecture (Figure 7.6) represents the overall working of the attendance management process. It is implemented in two modules:

1. Admin module
2. Faculty module

Both modules query the details from the admin and faculty, perform the attendance process, and generate the results. In the first module, the researcher created the student list using the MERN stack. The reason to implement this module is to perform the attendance process. Admin can manipulate the student details and also check the status of the student by date. Then the report of the attendance process is also generated by the admin and the report can be viewed by date.

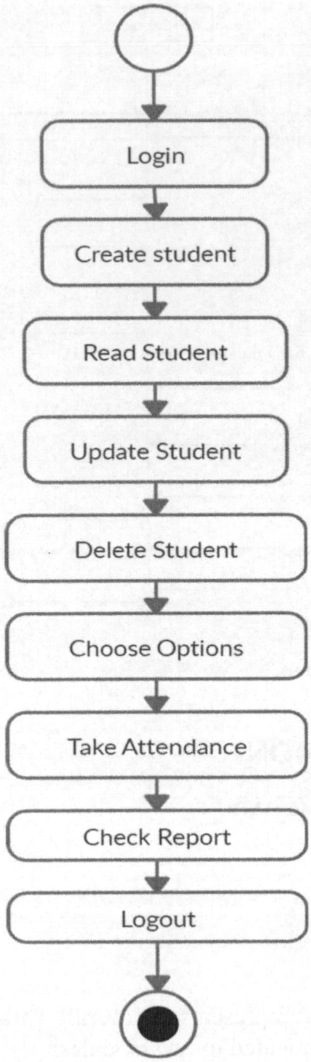


**FIGURE 7.5** Activity diagram.

In the second module, the researcher implements marking student attendance using the MERN stack. A user interface is created using Reacts and it is connected with the MongoDB database.

The dataset from the MongoDB database is displayed using the MongoDB Compass software on the faculty panel page. Then the attendance is marked by the

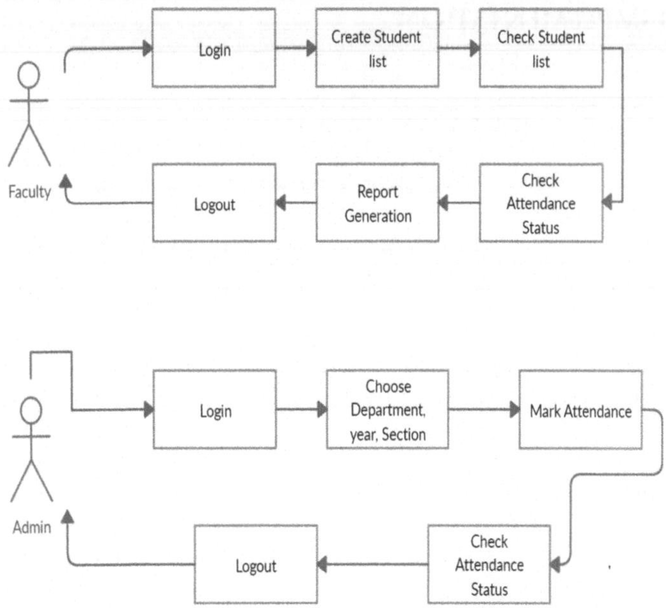


**FIGURE 7.6** Architecture diagram.

faculty, and the attendance data is stored in the attendance collection in the MongoDB database. Then the stored attendance is viewed as a status on the status page.

## Modules Description

### Faculty Module

In the faculty module, the web page is designed using the MERN stack. Details like name, gender, department, date of birth (DOB), year, and section are obtained from the created student list from the MongoDB.

The student list is displayed along with the date and the faculty chooses the particular department, year, and section to mark the attendance. Then the faculty is able to view the attendance status that they marked by the date.

### Admin Module

In the admin module, the web page is designed using the MERN stack. Details like name, gender, department, DOB, year, and section are stored in the MongoDB database. Only the admin can create the student list and manipulate the list. Faculty cannot modify the content of the student list. The status of the marked attendance is checked and maintained in the MongoDB database.

The report of the attendance is generated and can be downloaded in the CSV format.

## BACKEND IMPLEMENTATION

### INDEX.JS

```
//Middleware app.use(morgan('dev'));
//router
const router1 = require('./SchemaRoutes/StudentRouter'); app
.use('/',router1);
const router2 = require('./SchemaRoutes/AttendanceRouter');
app.use('/Attendance',router2);
//LocalHost
//DB Server Creation
mongoose.connect('mongodb+srv://mohanraj:5683263464@mern-app
- ilfch.mongodb.net/Attendance-
```

### Schema

*Student Schema*

```
Name: {
},
Regno : {
type: Number, required: true, trim: true,
min : 100000000000,
max : 999999999999
},
Department : {},
DOB : {
now, required:true
},
Year :{
type: Number, min: 1000,
max: 9999, required:true
},
Gender :
},
Section : {, trim : true
},
});
module.exports = mongoose.model('students', Students);
```

### Router Implementation

*Student Router*

```
}
});
//GET method with
}});
//BY id
({ "err": err });
```

```
}
});
//UPDATE method
router.put('/update', async (req, res) => { try {
const updStu = await students.update({ _ id: req.body. _ id },
{
$set: {
Name: req.body.Name, Regno: req.body.Regno,
Department: req.body.Department, DOB: req.body.DOB,
Year: req.body.Year, Gender: req.body.Gender, Section: req
.body.Section
}});
res.status(200).json(updStu);
} catch (err) {
res.json({ "err": err });}
});
//DELETE methodrouter.put('/:id',async(req,res) => { router
.delete('/delete/:id', async (req, res) => {
try {
const delStu = await students.remove({ _ id: req.params.id });
res.status(200).json(delStu); } catch (err) {
); }
});
```

## Front End Implementation

### Admin Panel

```
import { Modal, ModalHeader, ModalFooter, ModalBody, Table,
Spinner, Row, Col, For mGroup, Input } from 'reactstrap';
import Navbar from '../Navbar' import Axios from 'axios';
import '../../App.css';
import { MDBBtn, MDBIcon, MDBInput } from "mdbreact"; const
clearNullProp = (obj = {}) => {
for (const key in obj)
const getDateLabel = (date, format) => { let dateObj = new
Date(date)
if (format === 'slash') return dateObj.toLocaleDateString()
return dateObj.toDateString()}
const initFilterObj = { Department: '', Year: '',
Section: ''
}
class AdminDashboard
= {
//initial values modal: false, modal1: false, loading:
false, data: [],
editData: [],
```

```
_ id: "",
Name: "",
Regno: "", Department: "", Year: "",
DOB: "",
Gender: "",
Section: "", isEdit: false, mode: 'update',
filterObj: { ...initFilterObj }
}
this.isCreateClose = this.isCreateClose.bind(this) this.
infoChange = this.infoChange.bind(this) this.infoSubmit =
this.infoSubmit.bind(this) this.isCreate = this.isCreate.b
ind(this) }
isCreate() {
this.setState({ modal: true, mode: 'create' });
}
//modal update isUpdate = (data) => {
this.setState({ modal: true,
...data,
mode: 'update'
}) }
//modal close isCreateClose() {
this.setState({ modal: false,
_ id: "",
Name: "",
Regno: "", Department: "", Year: "",
DOB: "",
Gender: "",
Section: "",
});
}
//getting all values form DB componentDidMount() {
this.getAll();                          } infoChange = event => {
}) }
//Create and update functions infoSubmit = event => {
if (this.state. _ id === null) { let data = {
Name: this.state.Name, Regno: this.state.Regno,
Department: this.state.Department, DOB: this.state.DOB,
Year: this.state.Year, Gender: this.state.Gender, Section:
this.state.Section }
this.create(data)
} else {
let data = {
                                                }
this.create(data)                        } this.isCreateClose() }
//Create and Update API call create = data => {
if (!data. _ id) {
```

```
Axios.post("http://localhost:4000/", data).then(res => { this.
getAll()
}) }
else {
Axios.put("http://localhost:4000/update", data).then(res => {
this.getAll()
})                              } }
//GET all Function getAll() {
let { filterObj } = this.state
//                  clear null properties
filterObj = clearNullProp(filterObj)
Axios.get("http://localhost:4000/", { params: filterObj
}).then(res => { this.setState({
data: res.data
})                              }) }
//delete function del = data => {
var option = window.confirm(`Do You Want To Delete ${data.
Name}`) if (option) {
Axios.delete(`http://localhost:4000/delete/${data. _ id}`).
then(res => {
this.getAll();
})                              } }
viewButton() {
if (this.state.mode === 'update') { return (
<MDBBtn color="success" onClick={this.infoSubmit}>Update</M
DBBtn>)
}
else {
return (
<MDBBtn color="indigo" onClick={this.infoSubmit}>Create</M
DBBtn>)
}               }
modalHead() {
if (this.state.mode === 'update') { return (
<ModalHeader toggle={this.isCreateClose}>Update Students</
ModalHeader>
)                       }
else {
return (
<ModalHeader toggle={this.isCreateClose}>Creates Students</
ModalHeader>
)                             } }
// filter operations handleFilterReset = () => {
this.setState({ filterObj: { ...initFilterObj } }, this.
getAll) } handleFilterDepartmentFilterChange = (e) => {
const { value } = e.target let { filterObj } = this.state
```

```
filterObj = { ...filterObj, Department: value } this.set-
State({ filterObj }, this.getAll) }
handleFilterYearFilterChange = (e) => { const { value } =
e.target
let { filterObj } = this.state
filterObj = { ...filterObj, Year: value } this.setState({ fil-
terObj }, this.getAll)
                                        }
handleFilterSectionFilterChange = (e) => { const { value }
= e.target
let { filterObj } = this.state
filterObj = { ...filterObj, Section: value } this.setState({
filterObj }, this.getAll)
}
render() {
const { modal, loading, filterObj: { Department, Year,
Section } } = this.state; const isResetRequired =
Department|| Year|| Section
return (
<div>
<div style={{ marginBottom: '90px' }} >
<Navbar />
</div>
<h3 className="text-
center" style={{fontWeight:'bold', marginBottom:"30px"}}
>Students list</h3>
<Row style={{paddingLeft:"20px"}}>
<Col>
<MDBBtn color="cyan" onClick={this.isCreate}><MDBIcon
icon="plus- circle"/> Create</MDBBtn>
</Col>
<Col >
<FormGroup style={{marginTop:"0.6rem"}}> {/* col form group
margin
-top: 0.5rem; changesss */}
<Input type="select" value={Department} onChange={this.han-
dleFilter DepartmentFilterChange} name="departmentFilter"
id="departmentFilter">
<option value="" >Select Department
<option value="ARCHI" >Architechture</option>
</Input>
</FormGroup>
</Col>
<Col>
<FormGroup style={{marginTop:"0.6rem"}}>
```

```
<Input type="select" value={Year} onChange={this.han-
dleFilterYearFi lterChange} name="departmentFilter"
id="departmentFilter">
</Input>
</FormGroup>
</Col>
<Col>
<FormGroup style={{marginTop:"0.6rem"}}>
<Input type="select" value={Section} onChange={this.han-
dleFilterSect ionFilterChange} name="departmentFilter"
id="departmentFilter">
<option value="" >Select Section</option>
</FormGroup>
</Col>
<Col>
{isResetRequired && <MDBBtn color="deep- orange" onClick={t
his.handleFilterReset}>Reset</MDBBtn>}
</Col>
</Row>
<Table hover striped responsive borderless>
<thead className="thead-dark">
{loading && <Spinner color="success" />}
{this.state.data.length > 0 && !loading ? this.state.data
.map((e, i) => <tr key={e. _ id}>
<td>{i + 1}</td>
<td>{e.Name}</td>
<td>{e.Regno}</td>
<td>{e.Department}</td>
<td>{getDateLabel(e.DOB, 'slash')}</td>
<td>{e.Year}</td>
<td>{e.Section}</td>
<td>{e.Gender}</td>
<td><a color="warning" href="# " onClick={event => { this.
isUpdate(e) }}><MDBIcon className="icon" data-
toggle="tooltip" title="Edit" color="#xE147" icon="edit" /></
a></td>
<td><a color="danger" href="# " onClick={evevt => { this.d
el(e) }}>< MDBIcon className="icon1" data-toggle="tooltip"
title="Delete" far icon="trash-
alt" /></a></td>
</tr>) : <tr><td>{!loading && 'No results found'}</td></tr>
}
</tbody>
</Table>
<div>
```

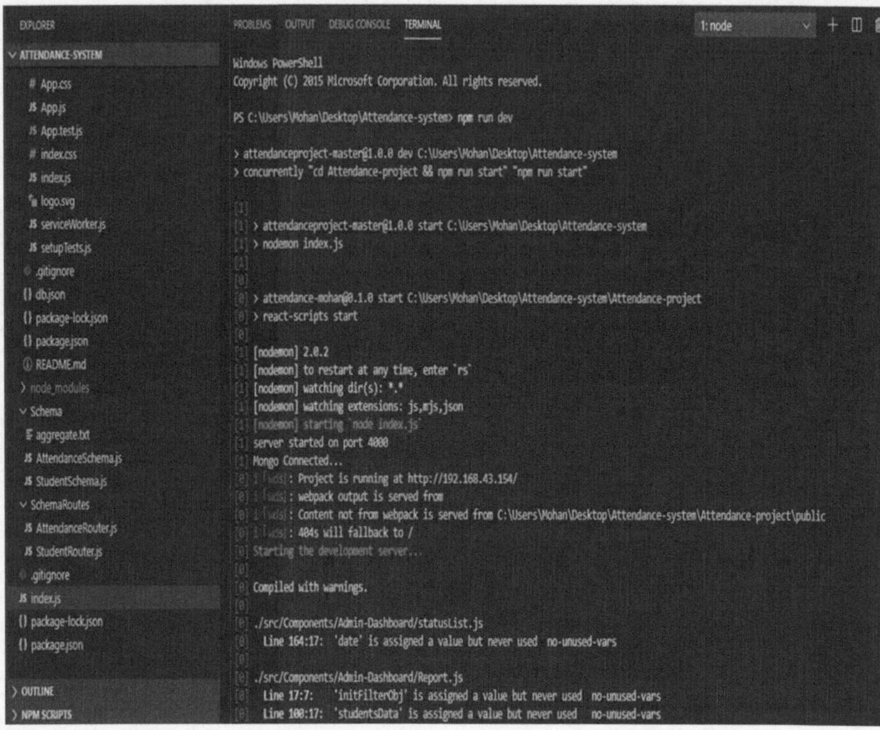


**FIGURE 7.7** Starting application.

### Backend Implementation

Figure 7.7 presents the snapshot of the application starting and Figure 7.8 shows the backend implementation using MERN stack.

### Front End Implementation

Figure 7.9 shows the user interface for the implementation of the attendance management system.

Figure 7.10 shows the attendance status and the submission of user details into the MongoDB database for the implementation of the attendance management system. Figure 7.11 shows the status list of the attendance system.

Figure 7.12 shows the report list of the attendance system.

Figure 7.13 shows the faculty dashboard of the attendance system.

## User Manual

### Instructions to Run the Project

- Install the Nodejs, MongoDB Compass, VSCode editor.

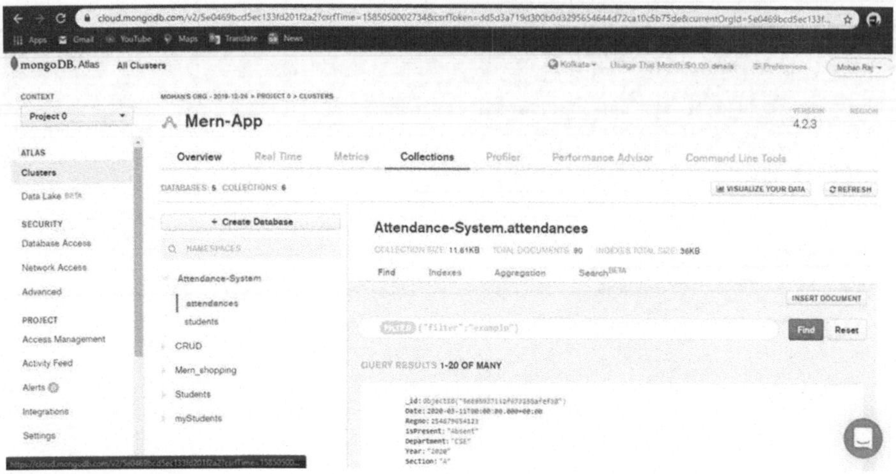


**FIGURE 7.8** Database.

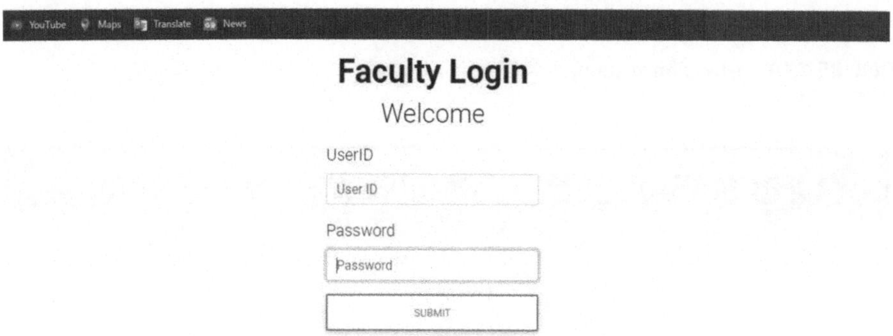


**FIGURE 7.9** Faculty login.

## Instructions for Node Analysis

- Open the VSCode editor.
- Open the project Folder.
- Open the terminal.
- Run the command "npm i" to install the required folders.
- Run the command "npm run dev" to run the project.

## List of Software

- NodeJS
- MongoDB Compass
- VSCode Editor

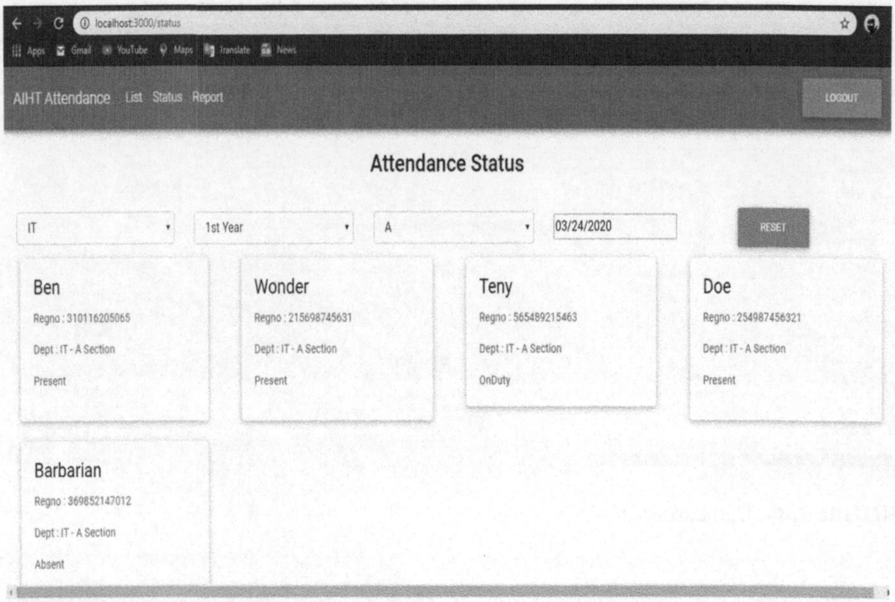


**FIGURE 7.10** Attendance status.

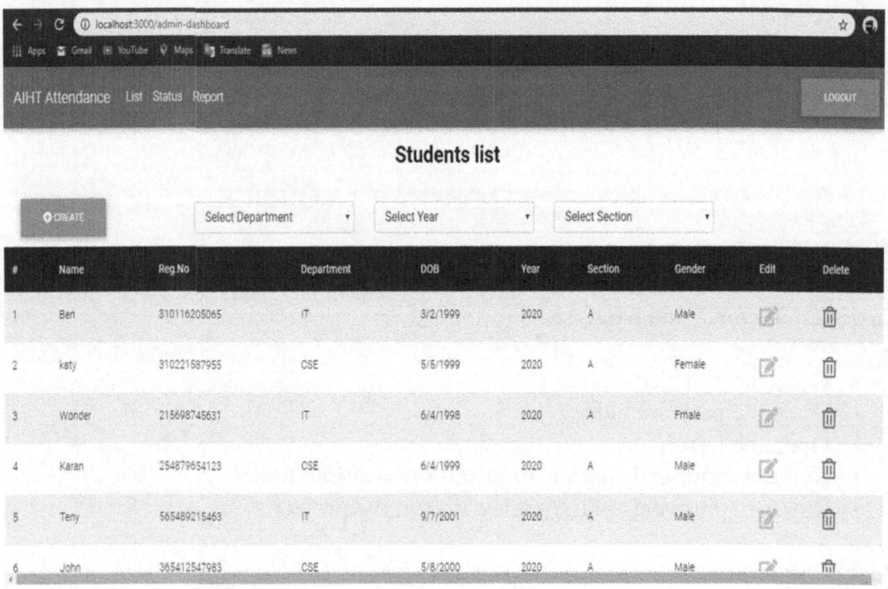


**FIGURE 7.11** Students list.

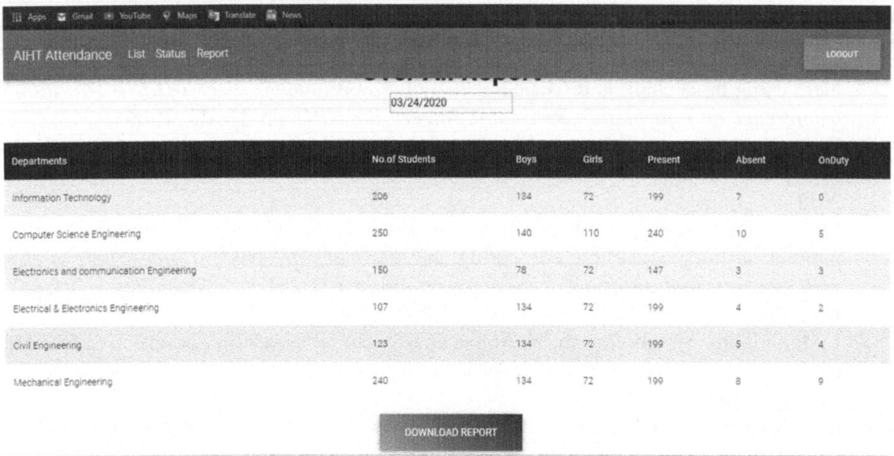


**FIGURE 7.12** Report list.

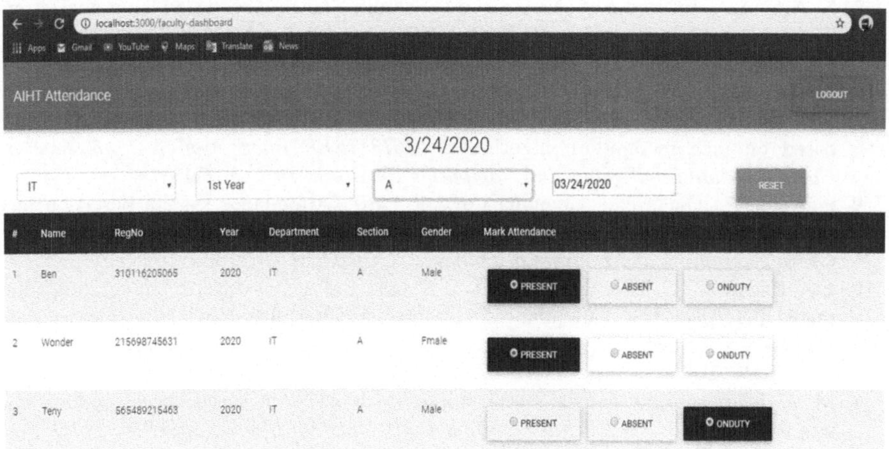


**FIGURE 7.13** Faculty dashboard.

## CONCLUSION AND FUTURE ENHANCEMENTS

The attendance of the students in a university or college and employees of an organization should be safely maintained and the database must be completely protected. For this purpose, blockchain is used. This system will be revolutionary in the future as it will be the safest nonrelational database to date. The retrieval of information and query processing is much easier. Though the researchers have incorporated the online attendance management system using the MERN stack with the basic attendance process, many future enhancements are required such as adding a specific login for each department to complete the attendance process.

## REFERENCES

1. G. Ayoade, V. Karande, L. Khan and K. Hamlen, "Decentralized IoT data management using blockchain and trusted execution environment", *2018 IEEE International Conference on Information Reuse and Integration (IRI)*, pp. 15–22, 2018.
2. E. Androulaki et al., "Hyperledger fabric: A distributed operating system for permissioned blockchains", *Proceedings of the Thirteenth EuroSys Conference*, pp. 1–15, 2018.
3. M. Aazam, I. Khan, A. A. Alsaffar and E.-N. Huh, "Cloud of things: Integrating internet of things and cloud computing and the issues involved", *Proceedings of 2014 11th International Bhurban Conference on Applied Sciences & Technology (IBCAST) Islamabad Pakistan 14th–18th January 2014*, pp. 414–419, 2014.
4. I. Makhdoom, M. Abolhasan, H. Abbas and W. Ni, "Blockchain's adoption in IoT: The challenges and a way forward", *Journal of Network and Computer Applications*, vol. 125, pp. 251–279, 2019.
5. G. Ramezan and C. Leung, "A blockchain-based contractual routing protocol for the internet of things using smart contracts", *Wireless Communications and Mobile Computing*, vol. 2018, 2018.
6. T. Lim, S. Sim and M. Mansor, "RFID based attendance system", *2009 IEEE Symposium on Industrial Electronics & Applications*, vol. 2, pp. 778–782, 2009.
7. N. Arbain, N. F. Nordin, N. M. Isa and S. Saaidin, "LAS: Web-based laboratory attendance system by integrating RFID-ARDUINO technology", *2014 2nd International Conference on Electrical Electronics and System Engineering (ICEESE)*, pp. 89–94, 2014.
8. S. Chintalapati and M. Raghunadh, "Automated attendance management system based on face recognition algorithms", *2013 IEEE International Conference on Computational Intelligence and Computing Research*, pp. 1–5, 2013.
9. P. Wagh, R. Thakare, J. Chaudhari and S. Patil, "Attendance system based on face recognition using eigen face and PCA algorithms", *2015 International Conference on Green Computing and Internet of Things (ICGCIoT)*, pp. 303–308, 2015.
10. J. A. Badejo, C. C. Eke, S. I. Popoola, T. O. Odu and A. A. Atayero, "Integrating automated fingerprint-based attendance into a university portal system", *2017 International Conference on Computational Science and Computational Intelligence (CSCI)*, pp. 1016–1020, 2017.
11. M. A. Islam and S. Madria, "A permissioned blockchain based access control system for IOT", *2019 IEEE International Conference on Blockchain (Blockchain)*, pp. 469–476, 14–17 July 2019.
12. H. Ardina and I. G. B.B. Nugraha, "Design of a blockchain-based employee attendance system", *2019 International Conference on ICT for Smart Society (ICISS)*, vol. 7, pp. 1–4, 2019.
13. J. Yang, Z. Lu and J. Wu, "Smart-toy-edge-computing-oriented data exchange based on blockchain", *Journal of Systems Architecture*, vol. 87, pp. 36–48, 2018.
14. Naresh Sammeta and Latha Parthiban, "Data ownership and secure medical data transmission using optimal multiple key-based homomorphic encryption with hyperledger blockchain", *International Journal of Image and Graphics*, vol. 23, no. 3, p. 2240003, 2023.
15. Naresh Sammeta and Latha Parthiban, "An optimal elliptic curve cryptography based encryption algorithm for blockchain-enabled medical image transmission," *Journal of Intelligent & Fuzzy Systems*, vol. 43, no. 6, pp. 8275–8287, 2022.

# *Section–III*

## *Blockchain Security in Healthcare*

# 8 Blockchain and AI in eHealth

*K P Revathi and T Manikandan*

## INTRODUCTION

Constant health monitoring of infants and children is used for evaluation of their vital parameters to avoid death. Noninvasive sensors are used to monitor the children. The caretaker or the parents can remotely monitor the child's vital parameters. Using a wireless sensor network (WSN), health parameters can be monitored by sending the data via a communication medium. Machine learning is used to analyze biomedical signals that are attained from the sensors to perform medical diagnosis. The machine learning approach detects the diagnosis with a human-like detection capability. The smart healthcare monitoring system is used to evaluate the physiological activities of the infant and child condition. These are used to monitor remotely in the home or hospital. To track the normal activity and everyday activity of the child, smart child healthcare monitoring is used. According to recent research, many researchers have developed flexible and biointegrated sensors to measure the physiological activity of children. These biointegrated sensors can stretch, bend, and fold into thin layers. They are small in size and are compatible with nature. They provide high-quality detection in measuring vital parameters. They are skin safe. They are soft, skin-like types with a high resolution of input. Different parameters that are measured are heart rate, pulse rate, lactose, blood pressure, pH level, emotional disorder, and hemoglobin level of the child and infant. The Internet of Things (IoT) is used to connect every node to another within the network. They are very accurate and used for real-time analysis. IoT plays a vital role in handling critical situations in healthcare. The behavioral condition and activity pattern of the child and infant are analyzed. These vital parameters can constantly monitor the health condition to avoid health deterioration. Some existing systems have many disadvantages and limitations in collecting, processing, and analyzing data. To avoid those limitations, IoT is used in the wireless media to track the physiological and behavioral analyses simultaneously. In this proposed model, the biointegrated sensors are used to collect the data of the child. A machine algorithm is used for gathering the information and to monitor the general status of their health condition. Tracking the health condition continuously helps to analyze health deterioration in the data records. Classifying the health condition of the child continuously can detect the abnormality in their health condition. Child health monitoring is one of the prominent factors to check

DOI: 10.1201/9781003428886-11

and monitor the constant growth of a child. The proper way of analyzing the behavioral and physiological parameters of a child for 0–3 years and 6–10 years of age is determined. Children's healthcare monitoring provides flexible monitoring of their health conditions. To monitor the physical activity using the framework, the proposed model is implemented. This framework provides access to the child's health data. A wearable sensor is a new development that is initiated to analyze the health of a child. In the traditional method, direct supervision is needed for analyzing the health condition of the child. By using this new methodology, the cost of a hospital and clinical laboratory can be reduced. It gives a routine check-up of the child's health and detects early intervention. Some of the early types of symptoms such as high fever, sweating, breathing difficulty, and pneumonia can be detected at an early stage. According to the researchers, viruses can attack the child by attacking the cells in their body. It can cause vulnerability to their cells. These viruses show an indication of fever. By detecting the body temperature abnormality, the hyperthermia can be analyzed. These non-invasive sensors are biodegradable and are safe and reliable (Figure 8.1).

The deaths of children sometimes occur due to sudden infant death syndrome. To reduce the premature death rate of infants and children, the continuous monitoring of the vital parameters is analyzed. Machine learning techniques are used for the computational purpose to predict health from the dataset. The combination of a smart system and machine learning will be the best approach to monitor the health of infants/children and to predict their conditions in health variations. A child who may be unable to understand their health conditions as they have low immunity and are unable to communicate with others. According to the World Health Organization (WHO), the death rate of infants and children is increasing due to the improper balance in their health conditions due to pneumonia and viruses. Constant monitoring, can identify health issues early. Discomfort occurs when the traditional way of monitoring in the infant and child is used. Modern biodegradable sensors are comfortable and can monitor vital parameters in the home or remote places. Hyperthermia can be detected early through abnormal variations in the body temperature. The

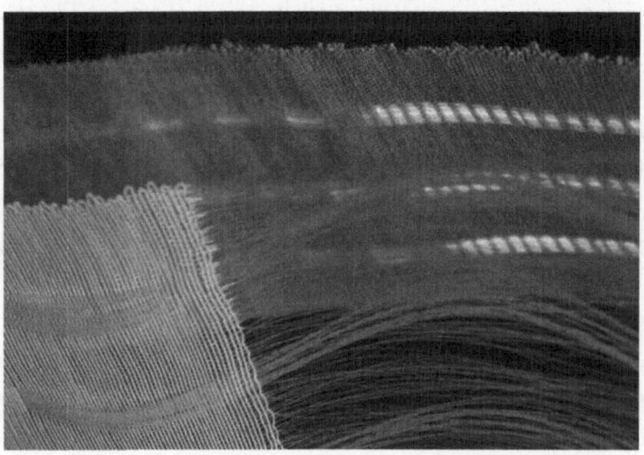

**FIGURE 8.1** Smart textile type biosensors.

dataset is recorded from the sensors, which are then analyzed, evaluated, and validated accordingly. In the recent approach, machine learning plays a prominent role in the prediction of early intervention using the optimal prediction analysis. Precise monitoring of different vital parameters is analyzed. Sensors are low in cost and are effective. Long-term monitoring can be ensured as they are compact and reliable to use. Everyday monitoring identifies the behavioral and physiological conditions of infants and children.

## LITERATURE REVIEW

Samira Shamsir et al. [1] wrote a paper that presents the development of multiple noninvasive sensors to measure physiological parameters using a smart infant monitoring system. They can evaluate the different vital parameters such as movement, noise, positioning, temperature, and humidity. They can also identify the ambient temperature. The acquired data are processed using the different levels of sensor nodes. In this proposed model, the system generates alarm signals when a deterioration in health occurs such as a seizure or apnea. These systems can detect the critical state of the respiratory system by measuring saturated blood level and breathing data. These data are analyzed using the machine learning approach. In this approach, the caretaker can monitor the data by checking the condition of the patient in a remote location through the wireless communication medium. These are used in clinical as well as in home-based environments. Shuai Xu et al. [2] proposed a paper where vulnerable patients who have neonatal cardiopulmonary can be detected. Detection of early medical intervention can help the death rate decrease. In the existing system, these physiological systems require multiple ranges of electrodes. These electrode wires are linked to which there is a wall-mounted display of units. In this research, the novelty is the use of wireless and biosensors to enhance monitoring by reducing generic injuries. These are soft, flexible, and biotype sensors that are easy and convenient to use. Souri et al. [3] introduced IoT to monitor the behavioral and physiological conditions of students. IoT is used for healthcare monitoring where it detects biological and behavioral parameters of the student. By using the experimental approach, the results are evaluated where they attain the proper accuracy of 99.1% in the range. Using different machine learning algorithms such as random forest, decision tree, and neural network, the results are determined.

Abdullah et al. [4] proposed a way to detect lower-back pain. A machine learning approach was used to assist diagnosing the disease at an early stage. The significant vital parameters are able to identify the abnormality in the spinal cord. The data are gathered using the unsupervised learning approach. Using principal component analysis and the supervised learning approach, the abnormality in the spinal cord region is determined. From the comparative analysis, the better performance is evaluated from the KNN classifier. The KNN classifier has a higher percentage when compared to the RF classifier approach. Fuentes et al. [5] defined four stages of child development during which safety and health are important. The child monitoring system emphasizes the benefits of monitoring health. The researchers provided the efficiency of the system requirements and different modules were represented.

Joseph et al. [6] proposed a paper about the body temperature of babies. The main aim of the paper was to ensure that the temperature is constantly monitored using the smart body temperature monitoring system. The temperature measures were taken periodically, especially during the nighttime to identify any irregularity in their temperature range. This project is a benefit for patients and caretakers as it reduces their anxiety. Once the body temperature exceeds the threshold range, the alarm is triggered.

Gupta et al. [7] initiated an efficient way of monitoring the health of infants using a wearable sensor system and wireless communication. Fully integrated sensors are used to measure the ECG, $CO_2$, and temperature of the infant. The wearable sensors set off an alarm or warning signs to identify the abnormality or sudden infant death syndrome. This system has some disadvantages:

- It does not support Arabic.
- There is no portable application for it.
- It lacks real-time live video and audio.
- It does not include cry identification.
- It doesn't support napping or being awakened.

IoT is always being improved, including its security. Waterproof ultrasonic obstacle sensors are used for sensing purposes, and these sensors are placed inside a simple locket that is given or worn by the baby. The locket will alert the parents or caregiver via mobile. Electrical devices are used for battery backup, and these devices use energy that is stored inside the caretaker's shoes and tracks the steps covered by the caretaker. First, the maximum and minimum values in each section of the signals curve are gathered. The frequently recorded maximum and minimum values are eliminated, and each peak is combined to ensure actual peak values.

The increasing number of crimes against children raises serious concerns about child safety and tracking. With this as a driving force, an intelligent IoT device for child safety and tracking was created to help the parents find and keep an eye on their kids. The system is created utilizing a single board that has embedded C-programming and interfaced with temperature, heartbeat, touch, GPS, GSM, and camera modules. The work is innovative in that when a child is in need of rapid attention during an emergency, the system instantly notifies the parent or caregiver by sending an SMS.

The child's sense of touch, body temperature, and heartbeat level are employed as parameters for the parametric analysis, and the results are presented. The aforementioned technique assures the safety and monitoring of children.

Ibrahim et al. [8] proposed a way to monitor the baby using Raspberry Pi sensors, which provide audio and video, and record the daily activities of the baby. It has cry detection and it supports the Arabic language. This proposed research was compared with the current system where it works in an efficient manner. Tupe et al. [9] wrote a paper on the methodology of a unique way of monitoring the healthcare of children. Using the framework, the physical condition of the child is progressed to reduce health vulnerabilities as the child grows. This framework provides information with flexibility and compatibility. Alam et al. [10] initiated a novel approach to classify the medical data using the classified approach. Feature ranking implements the medical dataset by using a ranking algorithm. The random forest algorithm predicts the features according to the high rank where it constructs the predictor. Using the different predictors approach, ten predictors were evaluated and performed using similar datasets. Ranking is performed using the tenfold cross-validated performance. Shreelatha et al. [11] presented a paper to provide a better way to protect the infant. Raspberry Pi is used for the advanced baby monitoring system. These monitor different parameters such as temperature, movement, and humidity. The camera is used to view the child by sending real-time video to their caretaker. In this approach, PIR sensors and sound sensors are used to measure the parameters. LEDs blink when the baby's crying and the system reminds caretakers of feeding and medication times.

Patil and Mhetre [12] initiated a GSM-based child screen framework. This framework detects important limits such as internal heat level, moisture status, heart rate, fetal development, and the use of GSM organization, and this information is sent to family members. Guardians can be alerted of boundary nuances so that any move can be made. A single microcontroller controls a set of sensors for detecting important parameters, an LCD display, a GSM connection point, and sound alerts.

This framework has a few drawbacks:

- An Arabic language support is not available.
- Neither the awake nor the sleep feature is supported.
- It is unable to detect humidity or room temperature.
- The cry detection feature is absent.
- There is no mobile application for it.
- It lacks real-time audio and video streaming.

Ferreira et al. [13] designed a wearable framework to monitor for sudden infant death syndrome, one of the significant reasons for death among newborn children during rest. The wearable IoT gadget is a remote sensor hub coordinated into a chest belt, and it screens for internal heat level, heart and breathing rates, and body position. In case of an emergency, the device will set off a visual and audible alarm and send a message to a mobile app.

Leier and Jervan [14] proposed a framework for checking children that can identify their primary vital signs and communicates the results to the control device, for example, a cell phone, over a remote connection. It is possible to use this framework in a variety of potentially life-threatening situations by monitoring the symptoms long term.

Suresh et al. [15] proposed the Advanced Baby Care System (ABCS) to help parents keep track of their child and their surroundings. The ABCS is equipped with an Ace Regulator (Arduino Mega 2560), which integrates all of the device's different modules by gathering vital signs from the sensor modules and sending messages to the trigger warnings.

Mothukuri and Prathap Kumar [16] developed a wireless Patient Monitoring System (PMS). The device measures heartbeat, temperature, and level of oxygen in the blood. The paper examines the plan and improvement of the low-cost device that utilizes GSM technology for observing patients.

The care of children is taken into very careful consideration, especially children who are very young, sick, and have severe restrictions on their ability to move around, communicate, and think. Additionally, children's immune systems are weak. Busy parents, a lack of parental understanding about health care, and a lack of parental awareness to monitor a child's health can all contribute to the lack of care, which can ultimately disrupt a child's growth, development, and healthcare process. The holistic healthcare framework that is suggested in this paper [16] may be customized to each child's needs and allows for the accurate, direct distribution of information to parents, caregivers, pediatricians, and hospitals. A smartphone belonging to a parent or nanny will be instantly provided with health care reminders and information. Parents can also dial a hospital or pediatrician with only one click in case of an emergency. To make it easier for parents and pediatricians to keep track of a child's health, all of the child's daily information—including information on their diet, activities, medications, and treatments—will be stored in an electronic medical record (EMR) and shown as a chart. Finally, the application of this holistic framework shields the child against numerous hazardous situations affecting their health, such as a medication problem or error regarding nutrition and exercise or inadequate healthcare.

Palaskar et al. [17] suggested a similar automated infant monitoring device. When crying is heard, a low-cost technology swings the cradle until the infant stops crying. The integrated alert sounds when one of the following occurs: the infant won't stop crying or the mattress is damp. Above the cradle is a video camera monitoring the child. However, parents are limited to receiving the system communications and cannot control it; they only receive SMS notifications. The system uses an IoT application to monitor the child and remotely operate the smart cradle in real time.

Harper and Blea [18] proposed an intensifier circuit for boosting the received sound signal as part of the child cry recognizer. The heartbeat generator circuit creates a heartbeat sign or pulse rate with 0 intersections that are lined up with 0 intersections of the amplified signal of sound as a result of the stronger sound signal. The beat signal is provided as an input to a signal affirmation circuit. The signal confirmation circuit produces a signal that indicates a baby's cry was heard.

Revathi and Manikandan [19] proposed a system that has an inbuilt alarm and GPS/GSM module that alerts users when a child crosses a physical boundary and needs assistance. By simultaneously calling and sending an SMS alarm, this method enables parents and other caregivers to take care of youngsters without having to physically attend to them. Parents can obtain the information from the system specified in the article to start the appropriate management measures, and it uses the cloud to monitor factors like baby cry, environment temperature, and moisture conditions. A system that monitors important characteristics like body temperature, heart rate, and moisture status was proposed by Wesseler et al. [20]. The specifics of the child's condition are communicated to the parents via GSM. The following equipment components are used: a temperature sensor, a moisture sensor, a pulse rate sensor, an

LCD display, a motion sensor, and a GSM module. The suggested framework featured a web application from which the sensor data could be shown remotely.

## RELATED WORK

### Internet of Things

The Internet of Things (IoT) is a network of objects connected to the internet. Devices may now transmit sensor data online without human interaction. IoT incorporates several devices and expands quickly due to its broad phase. According to predictions, there will be about 26.66 billion IoT devices operating by 2019 and 75 billion IoT devices available and wirelessly connected online by 2025. Millions of wearable sensors are among these connected gadgets, and they are extensively used in healthcare systems. IoT-related spending reached $737 billion globally in 2016 and was expected to reach $1.29 trillion by 2020. IoT is a fantastic industry with tremendous growth potential. Control, real-time monitoring, and independence or autonomous work and efficiency are all functions of IoT. IoT tries to make life easier, and people frequently invest in products that make their lives easier. This may be one of the primary reasons why IoT is so popular. As a result, the number of IoT applications is increasing across numerous industries. IoT was incorporated into the child monitoring system in this study to improve response times and provide parents a better sense of security.

The IoT's organizational structure as it relates to medical care applications aids in integrating the benefits of distributed computing and IoT innovation with the area of medicine. Additionally, it disperses the protocols for communicating patient data from various sensors and clinical devices to a specific medical care organization. The geography of a healthcare IoT (HIoT) is the overall framework or network of various IoT medical services that are well-connected in a healthcare environment. A fundamental HIoT framework consists of the distributor, the representative, and the endorser. The distributor speaks to a group of connected sensors and other medical devices that may or may not operate together to record the patient's vital information. Circulatory strain, pulse, temperature, oxygen saturation, ECG, EEG, and other measurements may be included in this information. The distributor might continually convey this data to a professional through the company. The dealer is responsible for managing and storing the collected data on the cloud. Finally, the supporter appreciates having constant access to and visual representations of the patient's data via a mobile device, computer, tablet, etc. When the distributor notices any physiological oddity or debasement in the patient's illness, they can handle this knowledge and offer feedback. A specific object is committed to each component of the IoT organization and cloud in the healthcare organization. It is difficult to suggest a widespread architecture for HIoT since the configuration depends on the interest in and use of medical care. A HIoT framework has historically embraced a number of fundamental modifications. While developing another IoT-based medical care framework for ongoing patient monitoring, it is crucial to drill down entirely relevant tests associated with the optimal well-being application. The IoT framework's outcome significantly depends on how items conform to the requirements of healthcare providers. The configuration should adhere to the clinical recommendations and processes in the decision-making process because every ailment necessitates a complex system of medical care exercises.

### Sensor

Sensors detect stimuli and change it into information. Normal sensors like the temperature sensor estimate the intensity of an item. Closeness sensors are used to identify the proximity of objects. Strain sensors ascertain the pressure applied. Optical sensors can detect light intensity. Humidity sensors will distinguish the presence of water vapor in the air. Miniature sensors are intended to gather and transfer data about the climate.

#### Cloud

Distributed computing implies shared registering assets (networks, servers, stockpiles, applications, and administrations) are conveyed over the web from cloud to client. As indicated, the cloud is an interconnected organization of servers offering types of assistance for individuals or organizations. The cloud upholds continuous, as a matter-of-fact activity, handling, investigating, associating, making due furthermore, and getting IoT gadgets along with applications. What's more, it decreases cost since clients are paying in light of utilization without building the physical foundation. It also permits engineers to make projects quicker. Associations can likewise get to big data from the cloud. The purpose of the cloud is to decrease handling trouble on clients. Therefore, various devices such as PCs, smartphones, and other tech-savvy phones are equipped to access different utility programs, storage, and application development platforms over the internet.

### Machine Learning in Healthcare

Healthcare use of artificial intelligence (AI) calculations helps in isolating perplexing and wide examples of information and records. For those with complex genomes and proteomics, this strategy is particularly well suited for clinical applications. Different issues are additionally recognized and distinguished utilizing it. Profound learning calculations are used in clinical innovations to create a patient treatment plan. Drug machine learning can be utilized in different ways, including:

(1) Infection detection and confirmation
(2) Customized change in treatment/lead
(3) Drug revelation/fabricate
(4) Audit of clinical test results
(5) Radiology and radiation treatment
(6) Health electronic records (HERs)
(7) Anticipating epidemics

## Machine Learning Algorithms in IoT

An astute locally situated remote medication box with an Android (healthcare IoT) application was assessed and detailed by the creators in request to further develop correspondence between patients and specialists. On the recommended stage, a smart medication box is accessible that advises patients when to take their prescription. The system provides brief notifications of medications prescribed to the patient's mobile through the Android application, with the added feature of remote web access. The patient gets the legitimate medication at the suitable time thanks to the machine's robotized warning. Moreover, the preset defender gets SMS warnings if any basic signs change. While the paper proposes a procedure or structure for observing patients' restorative use. It allows for the following of physician-recommended meds and directions. The framework recommends utilizing alerts to the patient. Clinical specialists frequently find infusions errors, and the e-health idea stops the disadvantages of customary gadgets. This smarter framework works better than traditional methods and is more modest, less expensive, more exact, and lighter. The recommended procedure could make it simpler for all senior people to take their prescriptions on time. As the world's population ages, there is more need to track down methods for helping people in their day-to-day existence. In such a manner, the Web of Things could give another aspect to current medical services by empowering a more individualized deterrent, and cooperative style of treatment.

This study advises professionals of an IoT for following and recording their patients' basic information while supporting crisis alert frameworks. The review proposes utilizing a wristband that might be connected to a cloud server to screen and help older people. It is expected to become a profitable remote systems administration arrangement. The qualities of a brain–computer interface utilizing different sensors, for example, electroencephalography (EEG), were examined in research. Envision utilizing a dissemination tensor (rsfMRI) to see and concentrate data from an epileptic patient's mind. For a methodology continuously utilizing advances to track, assess, and screen the brain, the proposed strategy incorporates frontline PCs. This guides in the early discovery and treatment of epilepsy.

The focus of this study is to foresee an "ictal beginning." Thorough study should be completed to oblige these novel requirements on the grounds that medical care frameworks need an organization structure that can empower QoS provisioning for video and time-delicate applications. Further, the article discusses the utilization of AI in medical services. Medical services will be influenced by AI in a couple of years. Later on, machine learning and artificial intelligence will change medical care, however, execution of their decision support systems (DSS) ought to give insight into issues looked at by patients and clinical experts. Large companies like Enclitic, Med Aware, and Google have made attempts at insight and innovation for the medical services industry. Successful medical services suppliers can't simply maintain for the time being. Man-made brainpower and AI will expand the effectiveness and precision. The sharing of such innovation could move along medical care administrations and lower medical services costs. IoT and medical care investigation are the significant focal points of the article. Big data analytics (BDA) is the aftereffect of the advancement of two related software engineering fields that cooperate to provide businesses information. Since a product is equipped for shipping, delivering, and trading involving a lot of unstructured, coordinated information, business experts make up the majority of big data clients. IoT increases the data capacity and trade across electronic and sensor-based devices. When making decisions humans consider their personalities, moral principles, information on the world, and political convictions. However, these components may be troublesome for information assortment used for learning. Machines are not predisposed to these hindering concerns. More client wellbeing will come about because of conquering machine learning and IoT imperfections. This paper also attempts to discuss the general viewpoint of IoT and machine learning that are utilized in the medical care framework, the applications utilized for customized medical care, and some other related works and their viewpoints with their discoveries of this field. It will additionally examine the difficulties and issues that confront computerized medical care.

### Wi-Fi Module

The Wi-Fi module is utilized for the process. Since it has Wi-Fi capabilities, the ESP8266 can either host an application or delegate all Wi-Fi networking tasks to another application. A newborn's health condition data will be automatically uploaded into the cloud storage from linked devices so the data is available everywhere in the world and is accessible to parents and doctors during examinations.

### GSM Modem

The term "GSM" stand for to Global System for Mobile Communications, which is used to establish a connection between a mobile device and a GSM modem. In order to enable connection with the network, GSM modems need a SIM (Subscriber Identity Module) card, a 12V power source, and a communication interface like RS232. The GSM modem uses the AT-Commands to facilitate communication.

## METHODOLOGY

### Textile Biosensors

Biosensors are highly flexible and user-interactive. They are user-friendly for the purpose of health monitoring. These devices can simplify the transmission of signals, minimize delay, and reduce noise. This wearable health monitoring device is comfortable on the skin and enhances sensing accuracy from the human body. These stretchable sensors are used to measure the heart rate by detecting the electrical activity in the heart. These elastic substrates have photodetectors to emit light from a source on the surface of the skin.

### Heart Rate

The volumetric ranges of variations of blood that is circulated all over the body are determined using heart rate (also see Table 8.1). Wearable sensors pay attention to

**TABLE 8.1**
**Vital Parameter Ranges**

| Parameter | Infant (0–3 years) | Infant (0–3 years ) | Child (6–10 years) | Child (6–10 years) | Electrode |
|---|---|---|---|---|---|
| Condition | Normal | Abnormal | Normal | Abnormal | Biosensors with textile sensors |
| Heart rate | 90–150 bpm | >160 bpm | 70–100 bpm | >140 bpm | Photodetector |
| Systolic blood pressure | 80–107 mm/Hg | >108 | 95–110 | >110 | Sphygmomanometer |
| Diastolic blood pressure | 55–70 | >70 | 60–73 | >73 | Sphygmomanometer |
| Pulse rate | 55–110 beats | >110 beats | 70–118 | >118 | Photodetector |
| Temperature sensor | 94.5–99.1°F | >100.1°F | 95.9–99.5°F | 100.4°F | Thermocouple/thermistors |
| pH level | 7.32–7.49 | >7.49 | 7.35–7.45 | >7.45 | Photodetector |
| Lactic acid | 0.5–2.2 mmol/L | >2.2 mmol/L | 4.5–19.8 mg/dl | >19.8 | Organic biosensor |
| Hemoglobin | 9.5–13 g/dL | >13 g/dL | 11.9–15.0 g/dL | >15 g/dL | Emitter and photo detector |
| Respiration rate | 30–60 bpm | >60 bpm | 20–30 bpm | >30 bpm | Organic photodiode |

the vital signals of the infant and child. They are small in size, low in cost, and consume less power. The electrical activity is measured from the P wave, T wave, and QRS complex. Traditional methods of using gel electrodes are harmful to children's skin. To avoid that, textile electrodes that are highly conductive and flexible are used. The reliable integration inside the clothing is convenient as it avoids direct contact with elements and children. Rather than using conventional gel electrodes, the textile electrodes are used where it is convenient and the signal quality is comparatively high with the accurate level of beat detection of the heart rate. The heart rate indicates the change in the emotional stability of the child or infant by their increase in the heart beat rate per minute.

## Body Temperature Sensor

The body temperature is measured using a thermoelectric effect that is noninvasive. These thermostats are embedded in clothing and accurately measure the skin temperature range. The temperature sensor, such as the LM35, is used with the textile sensor to attain the accurate value. Photodiodes are used to detect the reflected light from the skin. Minimum levels of light are observed during the contraction of the heart. Since the temperature varies according to the room temperature, the ambient temperature is validated.

## Blood Pressure Sensor

The blood saturation level is detected using a pressure sensor such as a sphygmomanometer sensor. They can analyze the data by the infrared reflection of the heartbeat rate. This device continuously measures the rate of the blood pressure by using the capacitive type of sensitive strain sensors. They are compressible and use robust electrodes. They use a piezoelectric capacitive strain sensor where it induces the generation of voltage. They are compressible orders made of plastic. The reflectivity activity of the data determines the systolic behavior of the heartbeat.

## Respiration Sensor and Pulse Oximetry

The organic photodiode is used to develop the flexible pulse range of the oximeter. Piezo-resistive sensors are used inside the clothing or fabric to attain the respiration rate of the child. During inhalation and exhalation, the indirect lung movement is attained by using the piezo-resistive sensors. When the blood level reduces, hypoxemia can occur. Different oxygenated blood is measured, such as tissue oxygenated level, venous oxygenated level, and peripheral oxygenated level.

## Hemoglobin

By using the photodiode, the volume and the concentration of hemoglobin in blood are determined by using the light absorption. Organic types of sensors are mounted inside, which consist of the photodiode and other sensors to detect the oxygenation level in the blood. These physical signal biosensors have motion detection to analyze the movement of the child. The ambient temperature is attained by using the LM35 sensor. The pH level and the lactic acid level are measured by using the biosensors. Biosensors are compatible and are less expensive. The hemoglobin level is used to identify any abnormality in the volume of the blood.

### PIR Motion Sensor

A passive infrared (PIR) sensor is basically made of a pyroelectric sensor that can distinguish levels of infrared radiation. A PIR sensor recognizes the infrared light emitted by a warm stimulus. It contains pyroelectric sensors that present changes in their temperature (in light of episode infrared radiation) into electric signals. Exactly when infrared light strikes a valuable stone, it creates an electrical charge. Likewise, a PIR sensor can be used to recognize the proximity of people inside an area of around 14 meters.

### Sound Sensor

One form of module utilized to hear the baby's sound is the sound sensor. Typically, the sound sensor outputs true or false values. The result is 1 if the sensor hears sound. If not, the value is 0. This allows us to determine if the infant is sleeping or awake.

### Ambient Temperature

The temperature in a room that measures the humidity in the air is called the ambient temperature. Along with the crucial metrics, the temperature and humidity of the room are measured. To check if surroundings are warm or cold to achieve a pleasant temperature state, it is important to take into account both the temperature and humidity. Heat-sensing devices are incorporated to measure the environment's ambient temperature.

## CONSTRUCTION

### Raw Data Collection

The raw data that are sent from the smart biosensors are integrated using the wireless communication. By using the wireless network, the raw signal is attenuated and collected in the cloud storage. The stored data are raw, unprocessed data. The unprocessed data have to be filtered and processed to avoid any signal distortion. Signal distortion has to be reduced to avoid the minimal loss. The child's data are analyzed according to their physical conditions. In general, the raw signal has noise distortion and random interference, which can increase the error percentage.

### Data Preprocessing

Raw data are processed to remove the noise distortion in the dataset values. By preprocessing the data, the format is getting changed in an effective way. The data are cleansed, encoded, and transformed into the known set of values by using the preprocessing technique. The unknown set of missing values data are disrupted accordingly. It changes the data into an understandable format for further use.

### Trained Dataset

The data are trained using the initial dataset and used to help the program. Each dataset is processed using the training model. The model can label and enrich the data by increasing the accuracy. To build a trained model, the set of information is processed. The model is tested and validated according to their typeset. For prediction purposes, an algorithm is used to train the model of the dataset. The data are organized according to their trained set of models.

### Data Annotation

Data annotation is used to add the metadata to the dataset. It is mainly used for labeling the dataset. It cleans up the data by annotating the dataset using a pipeline method. The machine learning approach is used to train the model of the dataset, so that the machine can be able to understand its format. It can make the machine easily understand the format by memorizing their input patterns.

### Data Validation

After training the data, the data gets validated before the transmission power. It checks whether the data is accurate by analyzing the dataset. The validated dataset is sent to the actual model to check whether it is free from error. The quality of the data is checked by validating the dataset. Each sample data is trained by using the training model by checking the null values or missing values. It removes the null values or the missing values to ensure that the data validation is complete. The preceding processes and flow are featured in Figure 8.2 and Figure 8.3.

## IMPLEMENTATION AND RESULTS

### Machine Learning Approach

Machine learning is used to classify and predict the data by labeling the data. Each data is labeled and unlabeled according to its specific conditions. Machine learning uses the known set of values to predict the result outcomes. The accuracy of the results is said to be the output. Machine learning algorithms classify the trained datasets. Machine learning is used to predict the disease by diagnosing it using the examination of both physiological and environmental factors. In machine learning,

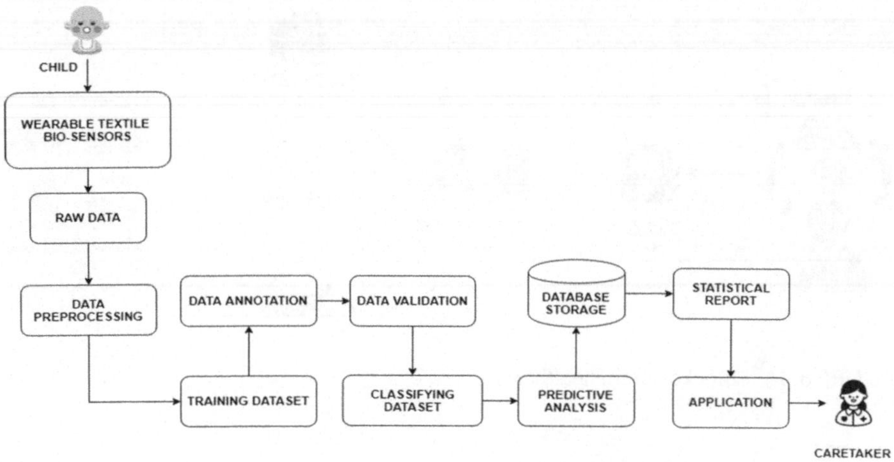


**FIGURE 8.2** Architecture diagram.

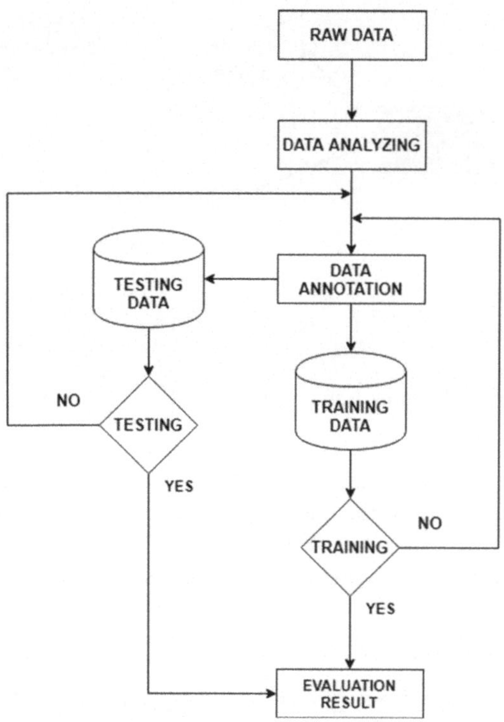


**FIGURE 8.3** Flowchart representation.

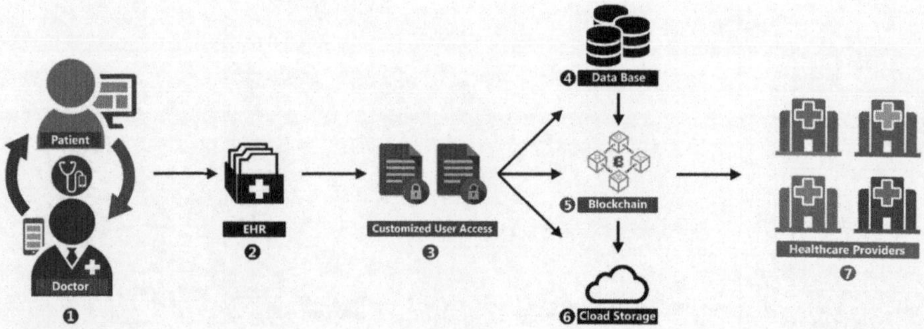


**FIGURE 8.4A** Blockchain in healthcare.

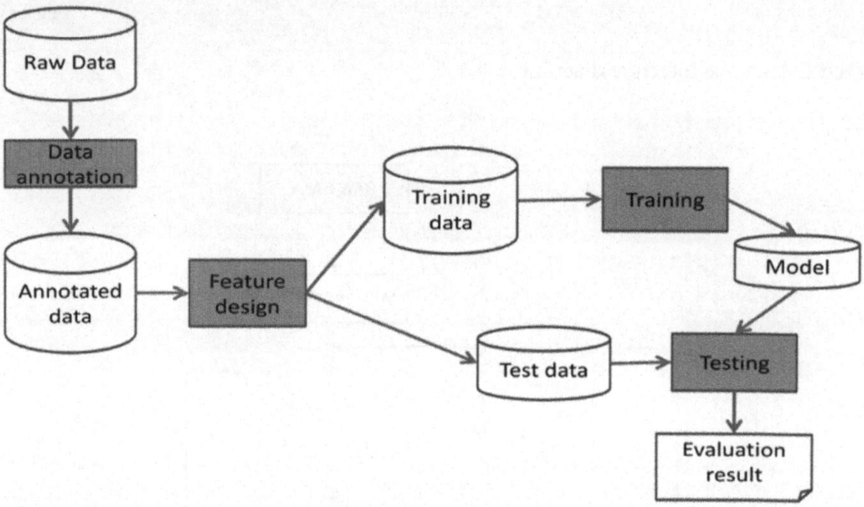


**FIGURE 8.4B** Machine learning framework.

the supervised learning approach is used to classify and predict the dataset as it emphasizes the patterns from the data. For distinguishing the feature, the output is predicted. It can validate the dataset by using the cross-validation technique. The machine learning framework is depicted with details in Figure 8.4b.

## Support Vector Machine

Support vector machine (SVM) is one of the types of supervised machine learning that is used to decide the boundary of multiple classes by performing various analyses of input within the dataset. It points at the position of the greatest possible distance between two data points in each class. It uses binary linear categorization as its main objective. It helps to highlight the model of a higher dimension. In the nonlinear model, it uses a kernel technique to map or plot within the high-dimensional space. The decision boundary between two different classes is using SVM. SVM supports every data point in the course entry of mapping the data. SVM can predict the distance between two data points where the scaling ranges from 0 to 1. For classification purposes, K-SVM is performed.

## Random Forest Classifier

Random forest is one of the significant methods used to integrate multiple models. The extensive ranges of datasets are classified with compatibility by using a random forest algorithm. Bootstrap aggregation and bagging models integrate multiple models of data. Model variants are reduced by using the bagging technique to prevent overfitting. Decision entries are used to specify the data point by attaining various outcomes in the decision tree. It was classified in the majority value by attaining the average value. In terms of prediction, a combined decision approach is better in performance than a single decision approach. Random samples are taken from the training dataset. It chooses the attributes from each training dataset and selects the data point from the tree. Random samples are taken from the training dataset. It chooses the attributes from each training dataset and selects the data point from the tree. These features are selected according to their ranking scores.

Step 1: Select a random set of K features from the total set of m features such as K<<m.
Step 2: From the K features calculate the node p by using the split point in the dataset.
Step 3: By using the split node, split the node p into daughter nodes.
Step 4: Repeat the process until the number of nodes is reached.
Step 5: Build a forest by using the number of ways (n) to create trees by repeating the same process.

## KNN Classifier

The KNN classifier is a popular machine learning approach used for classification and regression tasks. In this model, the unlabeled data points and the existing set of training dataset distances are determined in the featured set of stored space. It is highly compatible for predicting classes. The nearest set of data points are validated according to the K value of new observations. This KNN classifier can allocate the predicted classes according to class label volume. The optimized set of KNN algorithm generates a neighbor only with a positive relationship to the requester. It can evaluate the weight of the adjacent neighbor. The weight of the adjacent neighbor is determined by using the KNN algorithm by measuring the distance function. Different types of continuous distance function variables are used to weight the adjacent neighbor. By using the KNN classifier algorithm, the distance functions of the test and training data can be validated. The algorithmic steps to be followed are:

Step 1: Load the dataset of values.
Step 2: After loading the data, initialize the set of values of k.
Step 3: For prediction purposes, use the iterate set of values from 0 to the total range of the training set of data points.
Step 4: Euclidean distance d(x, y) is used to measure the distance metrics by calculating the distance between the training data and the test data.
Step 5: After calculating the data, sort the distance in ascending order.
Step 6: Gather the k set of rows from the sorted type of array.
Step 7: Gather the frequent class from each row.
Step 8: After the prediction, the process is returned to the original state.

## Blockchain in Healthcare

The distributed and safe method to collect and gather transactions is provided by blockchain technology. A blockchain stores information and offers advantages including privacy, interoperability, and audits of an immutable nature. It uses a digital ledger of transactions that distributes the networks throughout the computer system to duplicate the complete network. It creates a chain of immutable blocks using a decentralized, digital ledger of transactions. The public key and private key are used in the blockchain technology to verify validity. The storing of a child's health information is controlled by blockchains. By using digital transactions, the application notifies the health records. The decentralized system is used to eliminate the point of failure or delay in order to eliminate traffic flows. It can guarantee the information's scalability and robustness. The signature is used to identify the user's identification and to verify the data. The blockchain can be used to boost scalability by adding more nodes. Data storage in the cloud server is encrypted by using common cryptographic techniques. Given the sensitivity of health data records, the digital signature is authenticated in order to examine user data (Figure 8.4a).

## Comparative Analysis

Each dataset is validated by using the performance of the model. By using the different algorithms, the data are formatted where each data is normalized ranging from both the positive values and negative values. Some of them are overfitted and are validated to avoid any set of potential bias. In the training step process, the predicted value is tested. Different parameters are validated to attain the train and test values, such as positive value, precision, and recall, to validate the cross-validations. From the outcome of the ROC curve, the prediction is validated as it occurs from the test data values. Some of the data validations are calculated as follows:

$$\text{Sensitivity} = \frac{TP}{TP+FN}\text{; Specificity} = \frac{FP}{TN+FP}\text{; Precision} = \frac{TP}{TP+FP}\text{;}$$

$$\text{Recall} = \frac{TP}{TP+FN}\text{; Accuracy} = \frac{TN+TP}{TN+TP+FP+FN}$$

**TABLE 8.2**
**Attributes from Data Record**

| Attribute | Infant (0–3 years) | Child (6–10 years) | Number of Features | Training Samples | Testing Samples |
|---|---|---|---|---|---|
| Heart rate | 140 bpm | 110 bpm | 7 | 44 | 166 |
| Systolic blood pressure | 98 mm/Hg | 100 mm/Hg | 8 | 67 | 46 |
| Diastolic blood pressure | 65 | 70 | 9 | 24 | 67 |
| Pulse rate | 105 beats | 108 beats | 15 | 35 | 77 |
| Temperature sensor | 98.1°F | 99.1°F | 17 | 36 | 98 |
| pH level | 6.55 | 7.01 | 19 | 78 | 78 |
| Lactic acid | 1.7 mmol/L | 17 mmol/L | 21 | 43 | 54 |
| Hemoglobin | 10 g/dL | 13 g/dL | 34 | 68 | 49 |
| Respiration rate | 40 bpm | 20 bpm | 56 | 85 | 98 |

The dataset is from the Pima Indian research center. Nine attributes were taken from the children's vital parameters (Table 8.2). TP is the number of children who have an abnormality in their vital parameters, TN is the number of children who are healthy, FP is the number of children who have an abnormality in their vital condition, and FN is the number of children who are found to be normal. The features extracted are ranked according to their evaluation. The evaluated dataset uses each feature in the context according to the output variable. Based upon the Pima dataset from the medical records, the number of features is determined. Both the training dataset and testing dataset are combined for the predictive results. The feature extraction is used to reduce the number of features by using the dataset (Figure 8.5a). The dataset is created by using the new features from the existing features. The results of a comparison on the basis of accuracy and computational burden are shown in Figure 8.5b and Table 8.3.

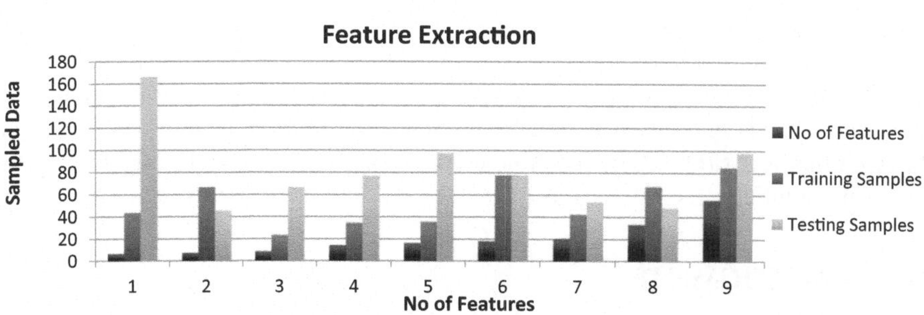


**FIGURE 8.5A** Feature extraction process.

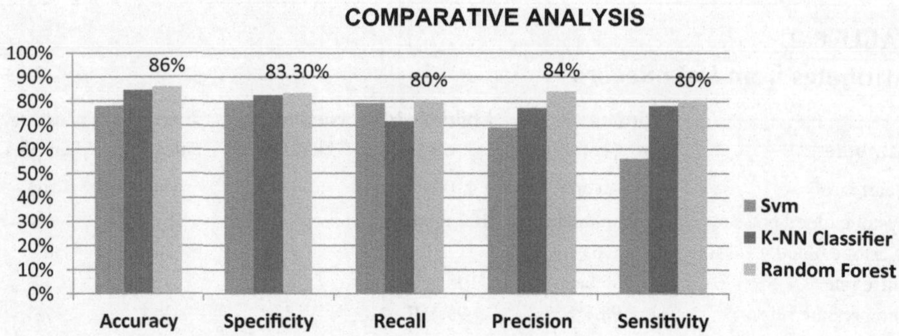


**FIGURE 8.5B** Comparative analysis of machine learning algorithms.

**TABLE 8.3**
**Comparative Analysis Performance**

| ML Algorithm | Accuracy | Specificity | Recall | Precision | Sensitivity |
|---|---|---|---|---|---|
| SVM | 78% | 80% | 79.0% | 69% | 56% |
| KNN classifier | 84.5% | 82.4% | 71.5% | 77% | 78% |
| Random forest | 86% | 83.3% | 80% | 84% | 80% |

## CONCLUSION

The chapter analyzes the smart ways of monitoring child health using noninvasive sensors. By using textile biosensors, the constant monitoring of health parameters is detected. Child health monitoring is one of the prominent factors to check and monitor the constant growth of a child. Constant monitoring of the system can detect any early abnormalities in their health conditions. A machine algorithm is used for gathering the information and to monitor the general status of their health conditions. These types of physical and smart biosensors are small in size, flexible, low in cost, and low in power consumption. Everyday monitoring identifies the behavioral and physiological conditions of an infant or child. The proper way of analyzing the behavioral and physiological parameters of a child 0–3 years and 6–10 years of age is determined. In this futuristic approach, the detection of ailments by analyzing the different diseases can be analyzed. By using the comparative approach over the dataset of the records, the random forest method has the highest accuracy at 86% than the two other algorithms. The machine learning approach is widely used in modern healthcare applications as it attains precise results and good performance accuracy.

## ACKNOWLEDGMENT

This work was supported by Supervisor T. Manikandan, Department of Electronics and Communication Engineering, Rajalakshmi Engineering College, Chennai, India.

## REFERENCES

1. S. Shamsir, O. Hassan and S. K. Islam, "Smart Infant-Monitoring System with Machine Learning Model to Detect Physiological Activities and Ambient Conditions," *2020 IEEE International Instrumentation and Measurement Technology Conference (I2MTC)*, 2020, pp. 1–6. https://doi.org/10.1109/I2MTC43012.2020.9129295.
2. Xu, Shuai, Alina Rwei, Bellington Vwalika, Maureen Chisembele, Jeffrey Stringer, Amy Ginsburg and John Rogers. Wireless skin sensors for physiological monitoring of infants in low-income and middle-income countries. *The Lancet Digital Health*. 3, 2021. https://doi.org/10.1016/S2589-7500(21)00001-7.
3. Souri, Alireza, Marwan Ghafour, Aram Ahmed, Fatemeh Safara, Ali Yamini and Mahdi Hoseyninezhad. A new machine learning-based healthcare monitoring model for student's condition diagnosis in Internet of Things environment. *Soft Computing*. 24, 2020. https://doi.org/10.1007/s00500-020-05003-6.
4. A. A. Abdullah, A. Yaakob and Z. Ibrahim. "Prediction of Spinal Abnormalities Using Machine Learning Techniques," *2018 International Conference on Computational Approach in Smart Systems Design and Applications (ICASSDA)*, 2018, pp. 1–6. https://doi.org/10.1109/ICASSDA.2018.8477622.
5. G. S. Fuentes and G. L. D. Intal, "E-healthcare: Child Monitoring Health System (CHMS) with SMS Functionality," *2020 IEEE 7th International Conference on Industrial Engineering and Applications (ICIEA)*, 2020, pp. 842–846. https://doi.org/10.1109/ICIEA49774.2020.9101976.
6. A. A. Joseph, T. H. L. Xuan, K. Kipli, K. L. Chin and N. S. Song. "Smart Body Temperature Monitoring system for Children via Mobile," *2018 International Conference on Computational Approach in Smart Systems Design and Applications (ICASSDA)*, 2018, pp. 1–6. https://doi.org/10.1109/ICASSDA.2018.8477613.
7. Sakshi Gupta, Zarnain M. Khan, Rupali Srivastava, P. A. Chougule. "Infant Monitoring System Using Multiple Sensors," *International Journal of Research in Engineering and Technology*, 2016. eISSN: 2319-1163 | pISSN: 2321-7308.
8. Ibrahim, Dina, Mohammad Ali Hammoudeh, Sadaf Ambreen and Sajid Mohammadi. Raspberry Pi-based smart infant monitoring system. *International Journal of Engineering Research and Technology*. 12, 2019, pp. 1723–1729.
9. Krishana Tupe, Vinod Jandrao, Sachin Chappe, Pawankumar Bhusalwad and A. V. Turukmane. "Child health care monitoring using sensor technology," *International Research Journal of Engineering and Technology (IRJET)*. 4(3), 2017, p-ISSN: 2395-0072.
10. Alam, Md, M. Rahman and Mohammad Rahman. "A Random Forest based predictor for medical data classification using feature ranking," *Informatics in Medicine Unlocked*. 15, 2019, p. 100180. https://doi.org/10.1016/j.imu.2019.100180.
11. Shreelatha, Shreya Pai, Sonal Cynthia Pereira, Tanya Nicole and Ms Ushadevi A., "Advanced baby monitor," *International Journal of Internet of Things*, 6(2), 2017, pp. 51–55. https://doi.org/10.5923/j.ijit.20170602.09.
12. S. P. Patil and M. R. Mhetre. "Intelligent baby monitoring system," *ITSI Transactions on Electrical and Electronics Engineering*, 2(1), 2014, pp. 11–16.
13. Andre G. Ferreira, D. Fernandes, Sérgio Branco, J. Monteiro, J. Cabral, A. Catarino and A. Rocha "A smart wearable system for sudden infant death syndrome monitoring," *IEEE International Conference on Industrial Technology (ICIT)*, 2016. https://doi.org/10.1109/ICIT.2016.7475060.
14. Mairo Leier and G. Jervan "Miniaturized wireless monitor for long-term monitoring of newborns," *14th Biennial Baltic Electronic Conference (BEC)*, 2014. https://doi.org/10.1109/BEC.2014.7320589.

15. Suresh, B., P. Ramaprasad, M. Mariyam and J. Abdul. "Advanced baby care system," *SSRG International Journal of Electronics and Communication Engineering*, 2(10), 2015, SSRGIJECE. ISSN: 2348 8549.
16. Chanakya Mothukuri and K. CH. Prathap Kumar, "Patient monitoring system," *International Journal of Science and Research*. 2(2), 2013, ISSN: 23197064.
17. R. Palaskar, S., A. Pandey, A. Telang, A. Wagh and R. M. Kagalkar, "An automatic monitoring and swing the baby cradle for infant care," *International Journal of Advanced Research in Computer and Communication Engineering*. 4(12), 2015, pp. 187–189.
18. Harper, M., and Maxine R. Blea. "Automatically rocking baby cradle," US3769641, Date: November. 6, 1973.
19. Revathi, K. P., T. Manikandan. "IOT based shrewd monitoring framework for children safety," *ECS Transactions*. 107(1), 2022, pp. 13967–13975.
20. Wesseler, M. E., R. Saatchi and D. Burke. "Child-friendly wireless remote health monitoring system," In *2014 9th International Symposium on Communication Systems, Networks & Digital Sign (CSNDSP)* (pp. 198–202). IEEE, 2014, July.
21. Aldahiri, Amani, Bashair Alrashed and Hussain Walayat. "Trends in using IoT with machine learning in health prediction system," *Forecasting*. 3, 2021, pp. 181–207. 10.3390/forecast3010012.
22. Sadhana, Tiwari and Sonali Agarwal. "Data stream management for CPS-based healthcare: A contemporary Review," In IETE Technical Review, 2021. https://doi.org/10.1080/02564602.2021.1950578.
23. Jeevan, Kharel and Haftu Tasew Reda. "Fog-computing based smart health monitoring system deploying LoRa wireless communication," In IETE Technical Review, 2019. https://doi.org/10.1080/02564602.2017.1406828.
24. Jabbar, W. A., H. K. Shang, S. N. I. S. Hamid, A. A. Almohammedi, R. M. Ramli and M. A. H. Ali. "IoT-BBMS: Internet of things-based baby monitoring system for smart cradle," *In IEEE Access*. 7, 2019, pp. 93791–93805. https://doi.org/10.1109/ACCESS.2019.2928481.
25. Uddin, Shahadat, Arif Khan, Md Hossain and Mohammad Ali Moni. "Comparing different supervised machine learning algorithms for disease prediction," *BMC Medical Informatics and Decision Making*. 19, 2019. https://doi.org/10.1186/s12911-019-1004-8.
26. Wei, Ding, Yukari Nagai, Liu Jing and Guo Xiao. "Designing comfortable smart clothing: For infants' health monitoring," *International Journal of Design Creativity and Innovation*. 7, 2018, pp. 1–13. https://doi.org/10.1080/21650349.2018.1428690.
27. Binu, P. K., V. Akhil and V. Mohan. "Smart and secure IOT based child behaviour and health monitoring system using hadoop," *2017 International Conference on Advances in Computing, Communications and Informatics (ICACCI)*, 2017, pp. 418–423. https://doi.org/10.1109/ICACCI.2017.8125876.
28. Ahmed, I. U., N. Hassan and H. Rashid. "Solar powered smart wearable health monitoring and tracking device based on GPS and GSM technology for children with autism," *2017 4th International Conference on Advances in Electrical Engineering (ICAEE)*, 2017, pp. 111–116. https://doi.org/10.1109/ICAEE.2017.8255337.
29. Zhu, Zhihua, Tao Liu, Guangyi Li, Tong Li and Yoshio Inoue. 2015. "Wearable sensor systems for infants," *Sensors*. 15, 2015, pp. 3721–3749. https://doi.org/10.3390/s150203721.
30. Revathi, K. P. and T. Manikandan. "A smart and secured approach for children's health monitoring using machine learning techniques enhancing data privacy," *IETE Journal of Research*. 69(3), 2023, pp. 1210–1221. https://doi.org/10.1080/03772063.2022.2150697

# 9 Blockchain-Driven Solutions for IoT Data Security in Healthcare

*P K Jawahar, S Mubeena,*
*V Jean Shilpa, R Anitha, and S Anusooya*

## INTRODUCTION

Recently, the trends and technology advancements of the Internet of Things (IoT) have made it possible for automation, data collection, and analytics in numerous industries. Communication between physical devices is made possible via the IoT, for example, using sensors for the collection and exchange of data with other devices over the internet. The hub connects various devices, gathers real-time data, analyzes it, and shares it locally. The applications of IoT include smart water and smart healthcare level monitoring, smart solar panels, smart home, smart waste management, and smart agriculture.

The Internet of Things makes it possible for the connectivity of inventory materials, over the Internet; information can be delivered or received. IoT has an impact on a variety of technologies, including embedded systems, artificial intelligence sensors, and real-time analysis. Smart devices can gather information and transmit it to perform specific, everyday life tasks. Smart cities, networked healthcare systems, intelligent transportation systems, and various other IoT applications are now accessible. IoT implementation in the medical sector is centered on a variety of sensors, medical devices, AI, diagnostic tools, and high-tech imaging equipment [1].

After the data has been connected to the control systems and collected from the sensors near or on patients, it is transmitted to the health monitoring unit. The cloud is used to store data, which helps manage the amount of data while preserving security. Since it is challenging to encrypt data from minimal-resource devices, security is a vital aspect of IoT. When data is sent from the sensor to the cloud, there is a chance that its integrity and confidentiality will be compromised. Due to its distributed nature, the cloud is the best option for storing health information that is simple for clinicians to gain access to while providing care for patients who reside far away. IoT and cloud use real-time processing, which increases complexity [2].

IoT is a bulk of networks that connect devices that are recognized, verified, and linked through cloud servers under the server–client paradigm, needing a huge amount of computing power and storing space. Additionally, communication

DOI: 10.1201/9781003428886-12

between the devices must happen over the internet regardless of how close they are to one another. Although this approach works well for small IoT systems, it cannot be scaled. It is expensive to set up a large number of communication connections, run centralized clouds, and connect the complete equipment for large-scale IoT networks [3].

## LITERATURE REVIEW

Sun et al. [3] propose storing medical data in the blockchain, which may be made secure and impenetrable by developing associated smart contracts. The plan leverages attribute-based access management to allow dynamic and granular access to that data. Additionally, IPFS technology is implemented into this system to decrease the storage-related issues on the blockchain. Tests show that the proposed system in this study ensures not only the integrity and safe storage of healthcare data but also increases efficiency when requesting access to that data. Prior to storing medical data in the blockchain, which may be made secure and impenetrable by developing associated smart contracts, the plan leverages attribute-based access management to allow dynamic and granular access to that data.

Ray et al. [1] focus on the Integration of blockchain and IoT technology as a logical option for creating a distributed IoT-based healthcare system. This paper discusses e-health and how it is used to assess popular blockchain consensus techniques. It assesses the blockchain platform's suitability for IoT-based electronic healthcare. Several practical application scenarios are provided to demonstrate how major IoT and blockchain capabilities could be used to assist healthcare services and ecosystems. It also provides a health data flow architecture, which unifies blockchain, and e-healthcare data can be stored, managed, and accessed through IoT.

Control over access to private medical data is possible using a blockchain technology data sharing permission paradigm. By using smart contracts, one can dynamically express consent about their health information and make the information searchable and available to data requesters. On the Ethereum blockchain, researchers put the idea into practice and evaluated several data-sharing scenarios. The model for individual permission for networks that share healthcare information has been developed as a result of the work by Jaiman and Urovi [4]. According to the assessment of the solution, such a data exchange architecture offers data requesters a flexible decision-making strategy regarding data utilization. The experimental evaluation demonstrates the efficacy of the suggested approach and its ability to adapt to customized access control settings in various data-sharing scenarios.

Gadekulla et al. [5] state decentralized, immutable, and traceable blockchain networks have the potential to reconfigure and revolutionize current IoT systems and provide them with improved degrees of security. A fresh paradigm known as BEoT, which combines the Edge of Things (EoT) with blockchain, was created specifically in conjunction with blockchain and IoT as a potential mediator for forthcoming services and applications. The researchers reviewed changes to BEoT technology and examined its immense potential in a range of application fields. BEoT is used in a range of commercial applications, such as smart grids, smart houses, smart towns, and smart healthcare.

Zghaibeh et al. [6] propose SHealth, which offers consumers maximum autonomy. It takes advantage of smart contracts by means of an easy-to-use graphical interface to start different requests and questions about, for example, consultations, diagnostic procedures, prescriptions, medical interventions, or past experiences with the patient. SHealth is easy to use, reliable, efficient, secure, and fully automated. The distributed database that stores the health-related data is accessible to all system participants without any sort of danger to the data's integrity. The paper discusses some scenarios involving the health system covered by SHealth.

Wu et al. [7] present a blockchain-enabled secure transaction network, a triple subject purpose-based access control (TS-PBAC) paradigm, and access control with diverse functions and responsibilities in the Internet of Medical Things (IoMT) scenarios. In an edge computation environment, the researchers created a local privacy differential (LDP)-based strategy and a role-based authentication mechanism to grant authorized users fine-granularity privileges in order to strengthen the security for important characteristics against an in-house attacker. The experimental findings demonstrate that the recommended privacy-preserving scheme can protect patient confidentiality in IoMT environments superior to conventional access control policies by making reliable and constant access control selections between information publishers and information requesters with different purposes.

## ARCHITECTURE OF IOT LAYERS

IoT in healthcare systems is essential for collecting, sharing, and transmitting data among devices in a smart way and with high speed [2, 8]. Sensors used in healthcare include the pulse oximeter, ECG sensor, temperature sensor, and sphygmomanometer. IoT uses several protocols like MQTT, HTTP, CoAP, TCP, and WebSocket for data transmission [9, 10]. The network connectivity used for IoT includes cellular technology up to 5G, Bluetooth, Wi-Fi, RFID, and low-power WAN. Smart healthcare devices collect a lot of data and these devices are also used in remote villages where people can self-monitor themselves and diagnose accordingly with doctors [11–13].

IoT can be classified into five different layers (Figure 9.1). They are the perception layer, which is used to transform analog signals to digital and digital signals to analog using sensors and actuators. The network layer is used to communicate between physical devices using some protocols like Wi-Fi, Ethernet, Bluetooth, Zigbee, MQTT, and HTTP [14]. The processing layer is used to capture, store, and process data for further requirements. The application layer is used to make decisions for business models once the data is analyzed. The business layer is used for business models with the processed data [3, 15, 16].

Perception layer: It is used for transforming analog signals to digital and digital signals to analog form. Sensors, actuators, and machine devices are used as application protocols [17–19]. Blood pressure monitors, working lights, and pulse oximeters are a few examples of medical equipment. Medical devices include oximeters, work

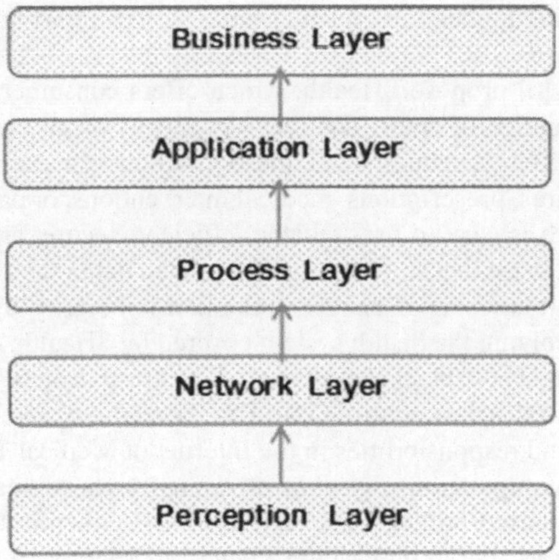


**FIGURE 9.1** Architecture of IoT layers.

lamps, and blood pressure monitors. Monitors, glucose meters, and other devices in the patient's room communicate with the IoT gateway using the aforementioned short-range links [20, 21].

Network layer: Communication takes place between the physical layer of IoT devices and IoT architecture. Wi-Fi, Zigbee, Bluetooth, MQTT, and Ethernet are some other application protocols used in this layer [22–24].

Processing layer: The process layer is used to capture, store, and process data for further requirements. CRM, ERP, and ERM are some application protocols used [25].

Application layer: Data is further processed and analyzed to gather business intelligence. Business decision-making software, artificial intelligence, and machine learning are some application protocols used [26]. In this layer, data is further processed and analyzed to gain business insight. Here, IoT systems connect to middleware or other programmers that can understand data more precisely [27].

Business layer: Business owners and other stakeholders make accurate future plans using data from the past and present. Data analysis is the application protocol used in this layer [28, 29].

Also sometimes included in the IoT architecture is the security layer, which follows security guidelines to install both firmware and hardware for IoT integration. Cloud security and connection security are important application protocols used. The importance of the security layer in IoT is crucial as IoT devices are often resource-constrained and vulnerable to cyberattacks, which can compromise the confidentiality, integrity, and availability of data, as well as the functionality of the devices themselves.

## APPLICATIONS OF IOT

The application of devices and sensors together with a range of technologies to acquire data from physical objects while interacting with other layers is one of the two primary uses for IoT devices. The complete network may be used with the process in connection with a supply-chain assessment that is proposed due to the RFID chip's ability to be connected to the product that needs monitoring and recognized with data that needs to be supplied. Realistically, the sector will benefit from the ability to consistently provide accurate information interchange using a variety of technological combinations.

Telemedicine is very popular and widely used in healthcare as a result of the growth of IoT technologies. A real-time tracking system for patients' vital signs and the condition of their rooms is proposed by a smart healthcare system inside the IoT environment. This system makes use of four pulse rate sensors, an electrocardiography sensor, a carbon dioxide sensor, and a carbon monoxide detector to get data from the hospital's environment. In each case, the error percentage of the developed scheme is within a predetermined range. Through a portal, medical staff can access patient status updates to process and assess the patient's current condition.

The primary objective of this chapter is to identify the impact and possibilities of IoT, blockchain, and fog computing in healthcare services. Blockchain technology has the ability to completely transform how we share, store, and transmit data by offering a safe and open platform for a variety of uses. The following text and Figure 9.2 describe the application of IoT in various fields.

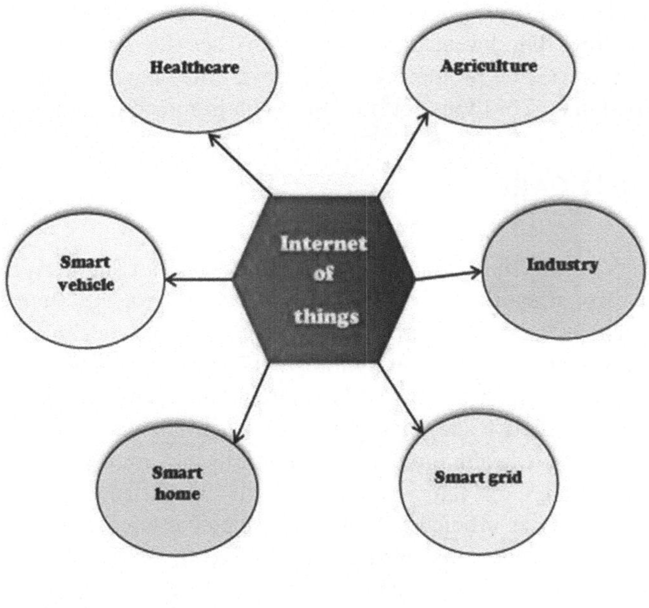


**FIGURE 9.2** Application of IoT.

Smart healthcare. Applications in healthcare include healthcare data management, telemedicine, remote access, and collaborative medications. IoT devices can be used to remotely check on patients' health, lowering the need for hospital visits and increasing patient outcomes. These gadgets have the capacity to gather information on vital signs, medication compliance, and other health factors and use it to guide treatment choices.

Smart agriculture. These are the applications to make smart farming, monitoring crop disease and pests, crop moisture and water levels, irrigation, and climate conditions. In order to optimize farming practices and raise yields, farmers can utilize IoT devices to monitor crop, soil, and weather conditions. These tools may gather information on a variety of parameters, including soil moisture, humidity, and temperature, and utilize that information to guide decisions for pest management, fertilization, and irrigation.

Smart industry: IoT devices can be used to monitor and manage commercial operations including production, delivery, and logistics. Industries are using IoT-powered production track and trace innovators and also intelligent robotics, reinvention, and warehousing. These gadgets can gather data in real time and use it to streamline operations, save downtime, and raise security.

Smart grid: Decision support software, improved management panel methods, and automation for power distribution faults and repairs. Precision sensing and measuring technology make it possible to have a smart grid.

Smart home: Lights may be turned on, temperatures can be changed, and doors can be locked all with the help of IoT devices. Homes can be made more convenient and energy-efficient by using these gadgets, which can be operated remotely using a smartphone or voice assistant.

Smart vehicle: Real-time data is gathered by IoT sensors on things like fuel use, engine temperature, fluid levels, and run time. IoT has the capacity to revolutionize a number of industries by promoting sustainable, efficiency, and safety while also enhancing the standard of living for both individuals and communities.

## IOT IN HEALTHCARE

1. IoT offers transformative applications in the healthcare industry, enhancing patient care, streamlining operations, and reducing costs. Remote monitoring enables healthcare providers to track patients' health in real time through wearable devices that monitor vital signs such as heart rate, blood pressure, and oxygen levels, while smart inhalers ensure timely medication usage. Predictive analytics, powered by IoT and machine learning, can analyze extensive health data to detect patterns and predict potential health issues, allowing early interventions for high-risk patients. In smart hospitals, IoT facilitates efficiency by locating medical equipment, automating tasks like medication dispensing, and managing supplies. Telemedicine, supported by IoT devices, extends healthcare services to remote areas, offering virtual consultations and reducing the need for travel. IoT also supports

personalized medicine by gathering data on patients' health and lifestyles, enabling tailored treatment plans. Furthermore, IoT combined with blockchain addresses key challenges such as secure data exchange and supply chain management. Blockchain ensures safe, decentralized, and tamper-proof health data sharing, empowering patients to control access while tracking medical supplies to prevent counterfeit drugs, thereby enhancing overall healthcare quality and safety.

## INTRODUCTION TO BLOCKCHAIN

Blockchain is the emerging technology used for data security in transactions. Each block includes an index number, transaction data, date, timestamp, and prior hash value. The previous hash function is the interconnection of the preceding block. The main features of blockchain technology include decentralization, transparency, security, increased capacity, authenticity, and immutability. Blockchain is classified into four different varieties, including consortium, hybrid, private, and public blockchains.

A distributed ledger technology, or blockchain, can be described as where data is stored in different blocks using cryptography hashing. Blockchain allows for a decentralized system that does not require centralization to transfer data to a specific organization. Blockchain monitors the exchange of data and maintains a distributed record of all data transactions across all nodes. Figure 9.3 depicts the basic concept of blockchain with the number of blocks connected consisting of the present block hash and previous block hash value.

Figure 9.4 depicts the structure of a blockchain with n number of blocks connected to previous the hash value.

A blockchain-based platform for developing and managing medical records and approvals is suggested. First, all healthcare institutions and clients have to finish the authentication function of the proposed scheme in order to receive both private and public key sets and register in the mobile application (MA). A distributed network is then used to run the suggested application. Before authorizing a user, the healthcare facilities must generate a medical certificate for any user who requests one. The authorities check the user information stored in the blockchain network using a

**FIGURE 9.3** Basic concept behind blockchain.

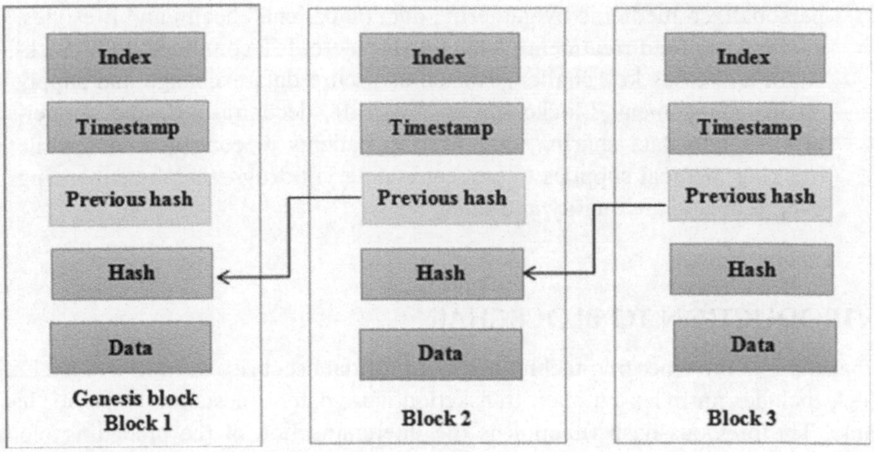


**FIGURE 9.4** Structure of blockchain.

mobile application interface. The blockchain network uses the data the user provided upon registration to create a proof of identity.

The decentralized aspect of the blockchain implies that there isn't just one entity in charge of the network, which makes it secure. Instead, the network's nodes cooperate to verify transactions and keep the ledger's integrity. The blockchain uses consensus algorithms like proof of work (PoW) or proof of stake (PoS) to make sure that each node concurs on the ledger's current state in order to guarantee the security and legitimacy of the transactions.

The transparency of blockchain technology, which enables all network users to view and confirm completed transactions, is one of its major advantages. Due to the fact that all transactions are recorded and made public to all parties, it is also harder for fraud to occur.

## Types of Blockchains

Blockchain is categorized into four types: public, private, hybrid, and consortium (Figure 9.5).

A public blockchain is a decentralized system that can be accessed by anyone with the help of an internet connection. An individual can perform a mining operation and has access to both existing records and new records.

A private blockchain is a partially decentralized system. Only one organization will have access and authority over the network. The private blockchain is also called a permissioned blockchain.

The hybrid blockchain is used by business and industry sectors that can control who can access their blockchain data and the information that is made available by establishing both a public system and a private, permission-based system, decentralized network (permissionless system). Within a hybrid blockchain, data and

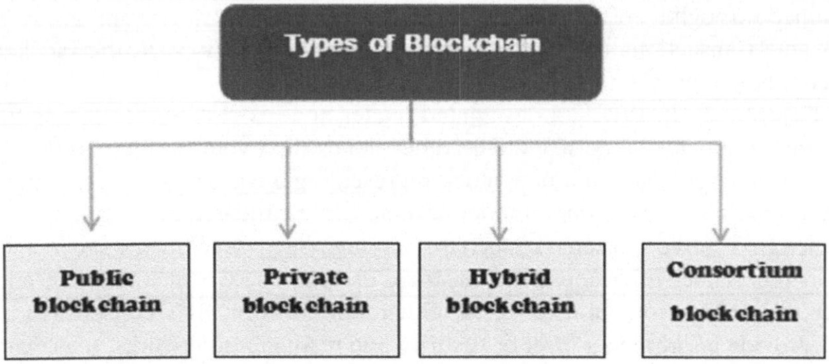


**FIGURE 9.5** Types of blockchain.

transactions are typically private, but consent shall be given through a smart contract if verification is necessary.

The consortium blockchain is a completely decentralized system. It gives access to multiple organizations and authority over the network. The nodes can send and receive transactions. A federated blockchain is another name for the consortium blockchain.

## BYZANTINE FAULT TOLERANCE ALGORITHM

A distributed network can achieve consensus (agreement on the same value) with the use of a feature known as Byzantine fault tolerance (BFT), which allows some nodes to answer incorrectly or fail to respond at all. The goal of using a BFT technique is to prevent faulty systems caused by utilizing group decision-making (including right and fault nodes), which seeks to lessen the impact of the malfunctioning nodes. The Byzantine generals' problem gave rise to BFT (Figure 9.6).

In a BFT-based blockchain network, a consensus protocol is used to ensure that all nodes agree on the validity of transactions and the state of the blockchain. This

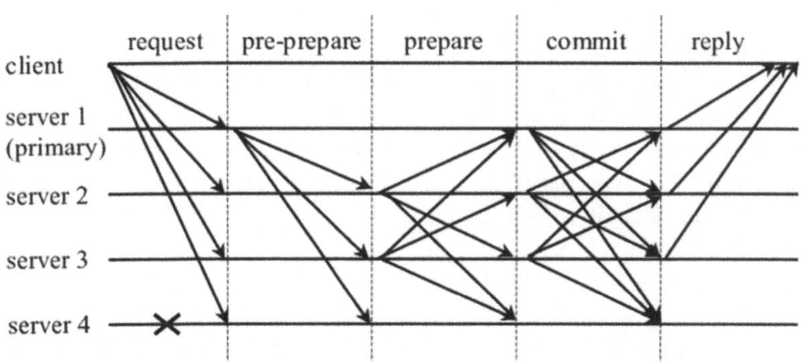


**FIGURE 9.6** Structure of Byzantine fault tolerance.

is attained through a voting process, where nodes in the network vote on the validity of new transactions and the ordering of those transactions in the blockchain. One common approach to BFT consensus is the practical Byzantine fault tolerance (PBFT) protocol. In PBFT, a designated leader node is responsible for proposing new transactions to the network. The other nodes then vote on the validity of the proposed transactions, and a new block representing a transaction is only added to the blockchain when a certain number of nodes have validated it.

Other BFT-based consensus mechanisms have been developed in recent years, such as the Tender mint consensus protocol used by the Cosmos network and the Clique consensus protocol used by the Ethereum network. BFT consensus mechanisms provide a significant level of security and resilience. Networks, making them suitable for use in applications where trust and reliability are critical. However, BFT-based consensus mechanisms can be computationally expensive and may require a high degree of network coordination, making them less suitable for use in smaller networks or applications with low transaction volumes.

Many ideas from the Byzantine error tolerance algorithm used in distributed systems are carried over into the practical version used in blockchain. In this instance, an agreement is reached to determine if a block is genuine. System nodes communicate with one another to add blocks to the chain. In this scenario, malicious nodes might communicate altered blocks, thus the network as a whole accepts the block as genuine if it is accepted by the maximum number of nodes [15].

Distributed systems are more prone to byzantine problems than centralized ones. These problems are a result of system node disinformation. The members of distributed systems are largely unaware of the causes of these errors or false information. It is so challenging to identify this node as malicious or flawed because it may be acting weirdly and providing various responses to various nodes in the network. Thus, to identify a faulty node, the system's honest nodes must reach an agreement. A mechanism that reaches a decision unaffected by a bad or malfunctioning node is referred to as a system with Byzantine fault tolerance [30].

The BFT mechanism consists of these three steps:

1. Pre-prepare: Each backup node receives a notification from the leader node via a prewritten message.
2. Prepare: Upon getting the pre-prepared response from the leader, the backup nodes reply to all the remaining nodes, even the leader, with the prepared message. A node cannot be regarded to be prepared until it has observed (2f + 1) prepared data from the other nodes and has acquired pre-prepared information from the leader.
3. Commit: The nodes send a message called a commit if they are prepared. If a node sees (f + 1) committed messages, it will carry out the client's request.

Two main categories of Byzantine failures are arbitrary node failure and fail-stop failure. Several illustrations of random node failures are:

- It does not yield a result.
- Draws an incorrect assumption.
- Responds with a purposely misleading result.
- Replies that various system components get have varying effects.

In a blockchain peer-to-peer (P2P) network, each node maintains a copy of the entire blockchain ledger, and each new transaction that occurs on the network is broadcast to all nodes in the network. This ensures that all nodes have the same version of the blockchain and that all nodes are aware of all transactions. When a new block is added to the blockchain, each node in the network verifies the transactions in the block and checks that the block has been added in a valid way according to the consensus rules of the network. If the block is valid, it is added to each node's copy of the blockchain, and the process continues. Because the network is decentralized, there is no central authority that controls the blockchain, and transactions can be validated without the need for intermediaries. This makes the blockchain P2P network secure and resilient, as it is not vulnerable to a single point of failure or attack.

### Steps to connect blockchain in a P2P network:

1. Consider the number of nodes to be connected in the network.
2. Each node acts as a client–server model.
3. Make an IP address for each node to be connected.
4. Search for the neo state command on the command prompt to find out the available port address.
5. Run the blockchain program to establish the P2P network on the edge server.

## PROPOSED WORK

A simulation test is made where a pseudo electronic health care data is generated and blockchain is applied to each patient's details. These secured details are connected in a distributed network using a P2P network. This P2P is implemented using the Python program. Each node is completely checked with the available port address for the edge server and the Byzantine fault tolerance algorithm is implemented in all the nodes. This is depicted as the model for the proposed work in Figure 9.7.

Figure 9.8 shows the simulation result of implementing blockchain to the patient health detail. In Figure 9.9 the results of node 2 without applying BFT are shown and Figure 9.10 shows the results of applying BFT in the P2P network. It is observed that when node 2 is connected to the P2P network it takes a minimum of 2 ms, maximum of 3 ms, and average of 2 ms to send data. When node 2 is connected using Byzantine fault tolerance algorithm the time taken to send data is a minimum of 1 ms, maximum of 3 ms, and average of 1 ms. So the transaction in BFT algorithm takes less time to commit the message. Results show that the blockchain transactions without the Byzantine fault tolerance algorithm in the P2P network take more time when compared with the Byzantine fault tolerance algorithm.

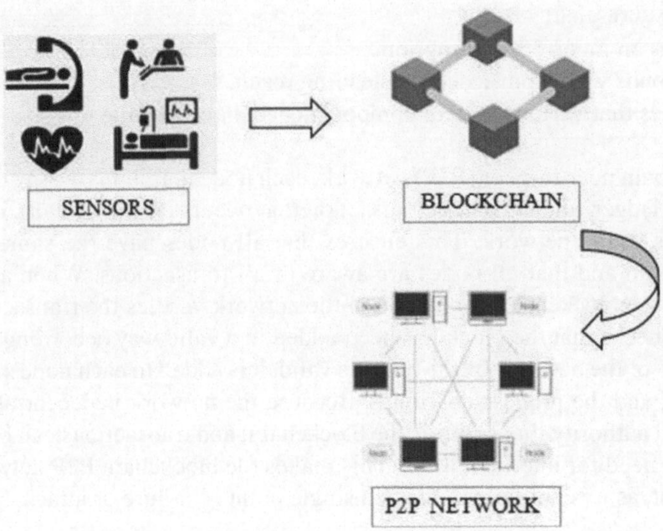


FIGURE 9.7 Proposed work.

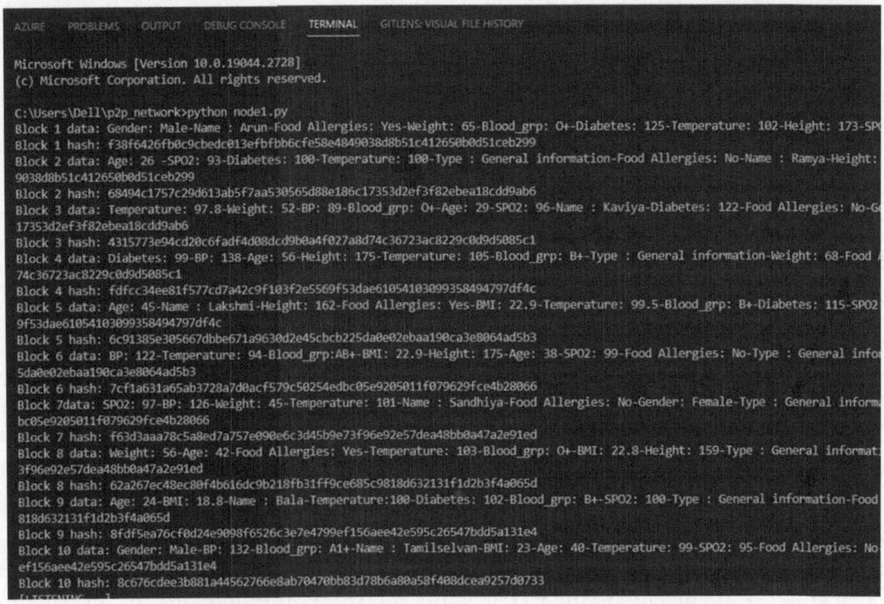


FIGURE 9.8 Implementation of blockchain for patient medical details.

The four phases of the Byzantine fault tolerance algorithm identify the fault node and resist that particular node. The remaining node prepares transactions and commits them with high speed. BFT-based blockchain networks can provide fast transaction processing and ensure the security and consistency of the network.

```
PS C:\Users\Dell\p2p_network> ping 172.20.108.222

Pinging 172.20.108.222 with 32 bytes of data:
Reply from 172.20.108.222: bytes=32 time=2ms TTL=128
Reply from 172.20.108.222: bytes=32 time=3ms TTL=128
Reply from 172.20.108.222: bytes=32 time=2ms TTL=128
Reply from 172.20.108.222: bytes=32 time=2ms TTL=128

Ping statistics for 172.20.108.222:
    Packets: Sent = 4, Received = 4, Lost = 0 (0% loss),
Approximate round trip times in milli-seconds:
    Minimum = 2ms, Maximum = 3ms, Average = 2ms
```

**FIGURE 9.9** Speed analysis of node 2 in P2P network.

```
PS C:\Users\Dell\pbft-1> ping 172.20.108.222

Pinging 172.20.108.222 with 32 bytes of data:
Reply from 172.20.108.222: bytes=32 time=3ms TTL=128
Reply from 172.20.108.222: bytes=32 time=2ms TTL=128
Reply from 172.20.108.222: bytes=32 time=1ms TTL=128
Reply from 172.20.108.222: bytes=32 time=1ms TTL=128

Ping statistics for 172.20.108.222:
    Packets: Sent = 4, Received = 4, Lost = 0 (0% loss),
Approximate round trip times in milli-seconds:
    Minimum = 1ms, Maximum = 3ms, Average = 1ms
```

**FIGURE 9.10** Speed analysis of node 2 using BFT algorithm.

## CONCLUSION

Many advancements in science and technology have changed the healthcare sector. Blockchain technology is implemented by using Byzantine fault tolerance, a consensus process for IoT data security. This project proposes a modified PBFT, in which the consensus process incorporates the mechanism for identifying fault nodes and resisting those nodes. The P2P network using BFT increases the consensus efficiency and performance speed. By using the BFT algorithm, blockchain transactions on IoT-based electronic health records are committed to the nodes connected in the distributed system and the speed of transactions between nodes is increased. Finally, electronic health record data are stored in edge servers so that the data storage problem can be reduced.

## REFERENCES


1. Ray, P.P., Dash, D., Salah, K. and Kumar, N., 2020. Blockchain for IoT-based healthcare: background, consensus, platforms, and use cases. *IEEE Systems Journal*, *15*(1), pp.85–94.
2. Dai, H.N., Zheng, Z. and Zhang, Y., 2019. Blockchain for internet of things: A survey. *IEEE Internet of Things Journal*, *6*(5), pp.8076–8094.
3. Sun, Z., Han, D., Li, D., Wang, X., Chang, C.C. and Wu, Z., 2022. A blockchain-based secure storage scheme for medical information. *EURASIP Journal on Wireless Communications and Networking*, *2022*(1), pp.1–25.
4. Jaiman, V. and Urovi, V., 2020. A consent model for blockchain-based health data sharing platforms. *IEEE access*, *8*, pp.143734–143745.
5. Gadekallu, T.R., Pham, Q.V., Nguyen, D.C., Maddikunta, P.K.R., Deepa, N., Prabadevi, B., Pathirana, P.N., Zhao, J. and Hwang, W.J., 2021. Blockchain for edge of things: applications, opportunities, and challenges. *IEEE Internet of Things Journal*, *9*(2), pp.964–988.
6. Zghaibeh, M., Farooq, U., Hasan, N.U. and Baig, I., 2020. SHealth: A blockchain-based health system with smart contracts capabilities. *IEEE Access*, *8*, pp.70030–70043.
7. Wu, G., Wang, S. and Ning, Z., 2021. Blockchain-enabled privacy-preserving access control for data publishing and sharing in the internet of medical things. *IEEE Internet of Things Journal*, *9*(11), pp.8091–8104.
8. Xiao, L., Ding, Y., Jiang, D., Huang, J., Wang, D., Li, J. and Poor, H.V., 2020. A reinforcement learning and blockchain-based trust mechanism for edge networks. *IEEE Transactions on Communications*, *68*(9), pp.5460–5470.
9. Ganesh, D., Seshadri, G., Sokkanarayanan, S., Bose, P., Rajan, S. and Sathiyanarayanan, M., 2020, October. Autoimpilo: Smart automated health machine using iot to improve telemedicine and telehealth. In *2020 International Conference on Smart Technologies in Computing, Electrical and Electronics (ICSTCEE)* (pp. 487–493). IEEE.
10. Saha, S., Sutrala, A.K., Das, A.K., Kumar, N. and Rodrigues, J.J., 2020, June. On the design of blockchain-based access control protocol for IoT- enabled healthcare applications. In *ICC 2020-2020 IEEE International Conference on Communications (ICC)* (pp. 1–6). IEEE.
11. Almakki, R., AlSuwaidan, L., Khan, S., Baig, A.R., Baseer, S. and Singh, M., 2022. Fault tolerance byzantine algorithm for lower overhead blockchain. *Security and Communication Networks*, *2022*. https://doi.org/10.1155/2022/1855238
12. Pang, Z., Yao, Y., Li, Q., Zhang, X. and Zhang, J., 2022. Electronic health records sharing model based on blockchain with checkable state PBFT consensus algorithm. *IEEE Access*, *10*, pp.87803–87815.
13. Rastogi, P., Singh, D. and Bedi, S.S., 2022, April. Design of a Blockcdhain based security algorithm for IoT in healthcare. In *2022 2nd International Conference on Advance Computing and Innovative Technologies in Engineering (ICACITE)* (pp. 2209–2214). IEEE.
14. Sara, A.B., Khadija, S., Abdessadek, A., Khalid, A. and Bouchaib, C., 2022. Impact of Blockchain technology in healthcare. In *ITM Web of Conferences* (Vol. 43, p. 01007). EDP Sciences.
15. Hegde, P. and Maddikunta, P.K.R., 2023. Secure PBFT consensus- based lightweight blockchain for healthcare application. *Applied Sciences*, *13*(6), p.3757.
16. Syed, T.A., Alzahrani, A., Jan, S., Siddiqui, M.S., Nadeem, A. and Alghamdi, T., 2019. A comparative analysis of blockchain architecture and its applications: Problems and recommendations. *IEEE Access*, *7*, pp.176838–176869.

17. Pelekoudas-Oikonomou, F., Zachos, G., Mantas, G., Ribeiro, J., Bastos, J.M.C. and Rodriguez, J., 2022, September. A scalable approach of practical Byzantine fault tolerance algorithms for IoMT blockchains. In *2022 IEEE International Mediterranean Conference on Communications and Networking (MeditCom)* (pp. 124–129). IEEE.
18. Shahnaz, A., Qamar, U. and Khalid, A., 2019. Using blockchain for electronic health records. *IEEE Access, 7*, pp.147782–147795.
19. Venkatesan, A.L. and Kumar, B.S., 2020, December. Role of Byzantine Fault Tolerance (BFT) in maintaining patient health records using block chain technology. In *2020 International Conference on Interdisciplinary Cyber Physical Systems (ICPS)* (pp. 1–6). IEEE.
20. Qu, J., 2022. Blockchain in medical informatics. *Journal of Industrial Information Integration, 25*, p.100258.
21. Jia, Q., 2021. Research on medical system based on blockchain technology. *Medicine, 100*(16), p.e25625.
22. Christo, M.S., Jesi, V.E., Priyadarsini, U., Anbarasu, V., Venugopal, H. and Karuppiah, M., 2021. Ensuring improved security in medical data using ecc and blockchain technology with edge devices. *Security and Communication Networks, 2021*, pp.1–13.
23. Namasudra, S., Sharma, P., Crespo, R.G. and Shanmuganathan, V., 2022. Blockchain-based medical certificate generation and verification for IoT- based healthcare systems. *IEEE Consumer Electronics Magazine.*
24. Arul, R., Alroobaea, R., Tariq, U., Almulihi, A.H., Alharithi, F.S. and Shoaib, U., 2021. IoT-enabled healthcare systems using block chain- dependent adaptable services. *Personal and Ubiquitous Computing*, pp.1–15. https://doi.org/10.1007/s00779-021-01584-7
25. Zivic, N., Ruland, C. and Ur-Rehman, O., 2019, April. Addressing byzantine fault tolerance in blockchain technology. In *2019 8th International Conference on Modeling Simulation and Applied Optimization (ICMSAO)* (pp. 1–5). IEEE.
26. Aujla, G.S. and Jindal, A., 2020. A decoupled blockchain approach for edge-envisioned IoT-based healthcare monitoring. *IEEE Journal on Selected Areas in Communications, 39*(2), pp.491–499.
27. Kassab, M., DeFranco, J., Malas, T., Laplante, P., Destefanis, G. and Neto, V.V.G., 2019. Exploring research in blockchain for healthcare and a roadmap for the future. *IEEE Transactions on Emerging Topics in Computing, 9*(4), pp.1835–1852.
28. Gunanidhi, G.S. and Krishnaveni, R., 2022, February. Improved security blockchain for IoT based healthcare monitoring system. In *2022 Second International Conference on Artificial Intelligence and Smart Energy (ICAIS)* (pp. 1244–1247). IEEE.
29. Ameta, G.K., Lavanya, P., Sudheer, S., Nagrath, S. and Chandramauli, A., 2022, April. Blockchain & IoT based smart home health monitoring system with a natural user interface. In *2022 3rd International Conference on Intelligent Engineering and Management (ICIEM)* (pp. 867–871). IEEE.
30. Meshcheryakov, Y., Melman, A., Evsutin, O., Morozov, V. and Koucheryavy, Y., 2021. On performance of PBFT blockchain consensus algorithm for IoT-applications with constrained devices. *IEEE Access, 9*, pp.80559–80570.

# 10 Integrating Blockchain in Healthcare to Secure Data

*Selvakumarasamy, Latha Parthiban, B. Hariharan, and R Parthiban*

## INTRODUCTION

Blockchain was first developed to supply distributed records of financial transactions that had been independent of central authorities or any financial institutions. Developments in blockchain technology have helped in improving exchanges that involve smart contracts, medical records, security, and allowing permanent access to data and insurance billing, along with that delivering a distributed database of transactions.

There are many advantages of blockchain of which the most important advantage in the medical records industry is that blockchain could also alter the exchanges between the databases used in healthcare, thus providing further access to medical records of patients, prescription and drug supply databases, tracking of devices, and hospital assets, comprising the whole life cycle of any device that is a part of the blockchain framework. Allowing access to the medical records of patients is very important to properly prescribe medication to patients, with blockchain being capable of histrionically improving the healthcare services infrastructure.

A blockchain comprises records that are arranged in a block structure in an ordered manner. Each data block comprises a hash that is a unique identifier (like a digital fingerprint), a hash of the previous block, and timestamps of recent transactions. Various blocks are linked in chronological order with this design, and the blocks that are connected are termed as a blockchain. It's virtually unfeasible to alter any of the blocks that are in the chain.

A smart contract is an algorithm that is executed automatically when the required conditions are satisfied, and it is a distinct unit from the already existing blockchain technology. The unit of smart contract was included in Blockchain 2.0 as a significant function, increasing the application and functions of blockchain beyond the field of cryptocurrency. The fact that blockchain cryptographically verifies the transactions executed does away with the possibility of a man-in-the-middle attack, replay, and all other traditional "device-to-cloud" attacks. With respect to this context, the aim of this research is to create a blockchain architecture for healthcare to secure data.

 DOI: 10.1201/9781003428886-13

## LITERATURE REVIEW

Previous research has highlighted the privacy and security concerns faced by health organizations and analyzed some of the solutions that could help solve those issues [1, 2]. The research by Agbo et al. [3] focuses on the challenges arising from patient-centered interoperability, which includes standards for data, privacy, and security, scalability, speed, incentives for blockchain, and blockchain/healthcare governance. Attaran [4] gives an overview of what Hyperledger fabric is, how it can be used to build solutions, and how transactions get executed in Hyperledger fabric.

Aste et al. [5] describe blockchain's concepts and provide a perspective on its risks and challenges. Other researchers have investigated privacy in electronic healthcare as a communication-centered issue since the future of electronic health systems will probably be quite distributed and will need the interoperability of a lot of subsystems [6–8]. In addition, the papers also describe the privacy needs related to research for persons other than patients. Others have also identified the IT-related privacy issues that occurred in the healthcare industry [9–11]. Tariq et al. [2] present a survey on the various challenges and direct issues encountered in security requirements and smart healthcare of the mentioned domains. They also discuss the capabilities of blockchain technology as a potential and strong security measure and analyze different blockchain-based approaches.

Previous studies provide a complete and detailed abstract on blockchain technology [12–16]. They provide an outline of blockchain architecture and compare it with some common consensus algorithms that are used in various blockchains. Efforts have been made to incorporate a lot of the compelling ideas from the community of professionals and researchers involved in improving the security and privacy of blockchain [17–21]. Scholars have also described how blockchain enables transactions and related data to be stored in a secure manner and verified without requiring any central authority [22–25].

## PROBLEM DESCRIPTION

It is quite important that healthcare systems share data in a systematic way. Also, they should be able to give privacy, increased access and control to the data, and guaranteed anonymity to the individual using the system. If there is a scope for less privacy, lower security, and trust handling, the person would be unwilling to give their sensitive and important information or even refrain from getting treatment. At present, a lot of healthcare data systems depend on a central party for managing confidential and sensitive healthcare data, which can be very susceptible to single-point-of-failure. However, due to its distributive nature, the technology of blockchain could alter this dependency. Blockchain can present the ability to solve technical issues and system attacks in an immutable and distributive way. Along with this, a record is included to track the ownership and authenticity of data.

## Existing System

Before the advancement of present-day technology, the healthcare industry was dependent on a system based on handwritten, paper medical records. This paper-based system of medical records was unorganized, inefficient, and vulnerable to tampering. Also, the system encountered the problem of redundancy and duplication of data since all the hospitals that the patient went to had different copies of his or her medical records. The healthcare sector then shifted to electronic health records (EHRs). These record systems were then used to store laboratory-based results and clinical observations. It was recommended that to strengthen the safety of the data of the patients, breaches be prevented and access to the stored information be increased. The main aim of EHR systems had been to fix the issues faced by the healthcare system that was dependent on paper to build a reliable system that could revolutionize the healthcare industry. They faced the problems of being unreliable, less secure, lacking interoperability among hospitals, and less user-friendly. Given that a lot of these EHR systems are used in different institutions, it was found that the systems had no universally defined standard.

## Proposed System

The primary aim of this research is to integrate blockchain with healthcare to create a system to ensure that patients' data is accessed only by authorized persons and isn't changed by a third party for malicious reasons. A blockchain network will be configured to securely store the collected data using a blockchain framework platform. The patients and healthcare workers will be a part of the blockchain network deployed as accounts having access to it and data will be stored after being considered as valid by the blockchain consensus mechanism. A Graphical User Interface (GUI) will be designed that allows an authorized user within the network (doctors, nurses, patients) to view patients' records.

## Requirement Gathering

The collected data is secured after every transaction in the blockchain for healthcare devices. One of the main sources from which the data will be collected is from patients' healthcare device readings. More precisely, data collected from patient-connected devices will be sent to the blockchain system as a transaction. If the transaction is considered valid by a node in the blockchain network, then it will be added as a block to the network. All nodes (patient/doctor computers) that are part of the network will also be updated with the new data.

## Analysis of Requirements

### Functional Requirements

- Allow users or patients to register with their Ethereum address.
- Allow patient to give permission to doctors to edit their records.

- Allow doctors to enter medical reports for patients and attach their digital signature with it.
- The patient can view and share their records. No unauthorized person can view the records.

### Nonfunctional Requirements

- Efficiency requirement: The system should use the available resources and dataset efficiently to produce fast and accurate results.
- Reliability requirement: The software system should be a reliable and easier to handle for both owners and customers. Since it is a very sensitive application, the software system should be reliable and not fail in any case. There must be recovery protocols established in order to avoid such issues.
- Implementation requirement: Implementation of the system is done using JavaScript and solidity, which makes it versatile and flexible to be used in any type of work environment.
- Delivery requirement: The complete system is envisioned to be delivered within four to five months, including an evaluation by the project guide every week.

### Data Source

The data for our project was collected from the internet.

### Cost Estimation

The researchers will be using an organic type COCOMO model for approximate cost estimation: KLOC = Thousand lines of code (estimate) = 2

- Effort undertaken = E = a/(KLOC)^b = 2.4(2)^1.05 = 4.969 person-months.
- Time required = D = c(E)^d = 2.5(4.969)^0.38 = 4.597 months.
- Staff size = ss = E/D = 4.969/4.597 = 1.08 (roughly 1–2 persons).

## SYSTEM ARCHITECTURE

Figure 10.1 presents the overall blockchain system. The diagram provides an overall physical arrangement of the blockchain model of the researchers. The diagram shows how a healthcare worker creates a medical record in the application, how MetaMask verifies transactions by generating digital signatures, and how patients can view their medical records from the application.

### Interface Prototyping (UI)

Figure 10.2 shows the user interface prototype of the work.

Figure 10.3 shows the interface to create a new address.

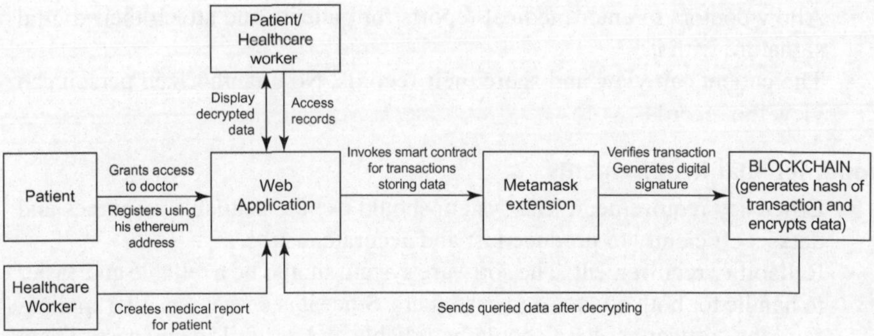


**FIGURE 10.1** Overall blockchain system.

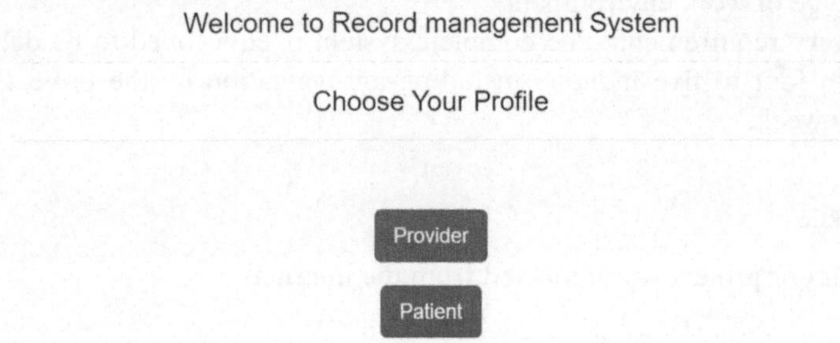


**FIGURE 10.2** Interface prototype.

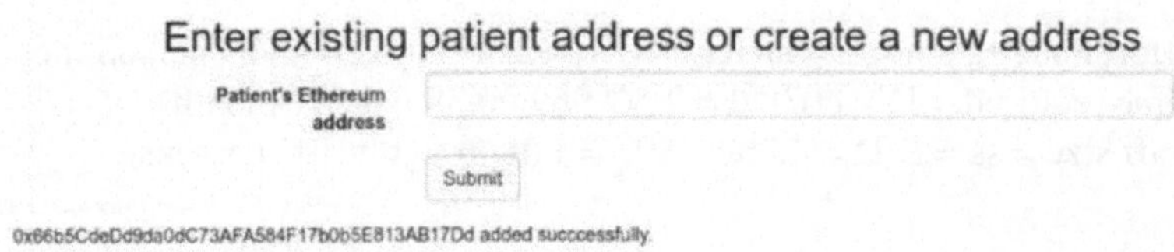


**FIGURE 10.3** Create new address.

Figure 10.4 shows the interface to create a new record and Figure 10.5 shows how the record is shared.

Figure 10.6 shows the GUI for creating permissions.

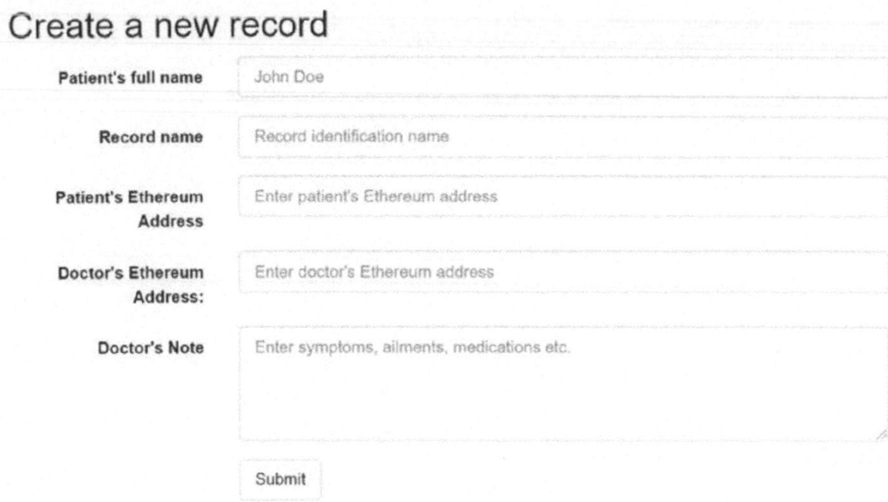


**FIGURE 10.4** Create a new record.

**FIGURE 10.5** Share record.

## Data Flow Design

Figure 10.7 shows the processes for sending data from the input data store to reports. The data created by patients and healthcare workers are stored in the blockchain and they can retrieve the data on request.

## Use Case Diagram

Figure 10.8 shows the user's interaction with the system. The users in this system are patients and healthcare workers. The diagram shows how a patient registers, gives permission for the doctor to add a patient record, and how a patient can view their record through the blockchain.

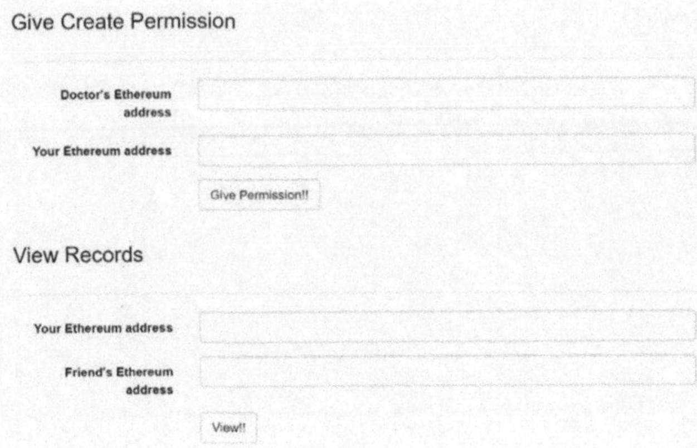


**FIGURE 10.6** Create permission.

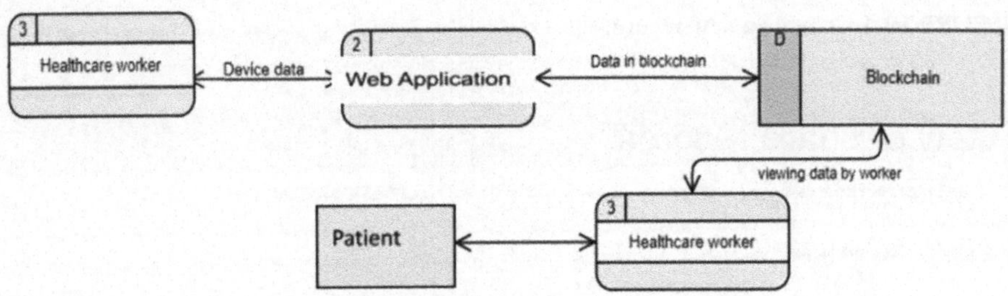


**FIGURE 10.7** Complete data flow design.

## Class Diagram of the System

Figure 10.9a shows how objects are related to one other and the operations taken. Each patient record is stored in the blockchain, the healthcare worker can add patient data, and numerous patient data is stored in the blockchain.

## Activity Diagram

Figure 10.9b shows various decision paths undergone during system activity. The activity starts with the patient registration, followed by logging in as either a doctor or patient, and ends with the patient viewing their records.

## Sequence Diagram

Figure 10.10 shows how the model works. Patients and healthcare workers are characters in the diagram. The model shows how a patient registers, views the record

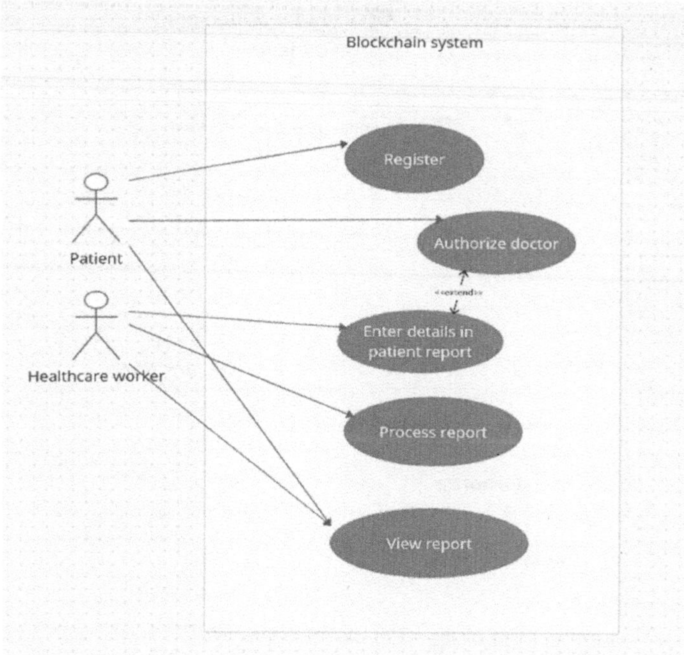


**FIGURE 10.8** Interaction with system.

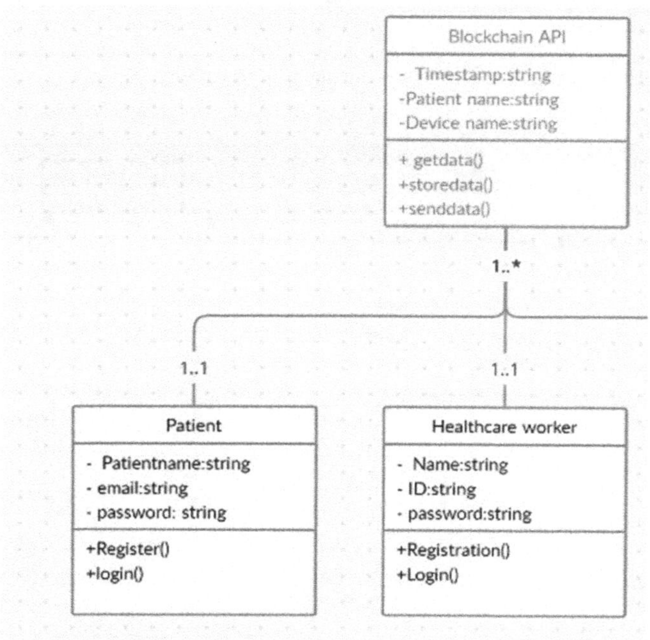


**FIGURE 10.9A** Class object and links.

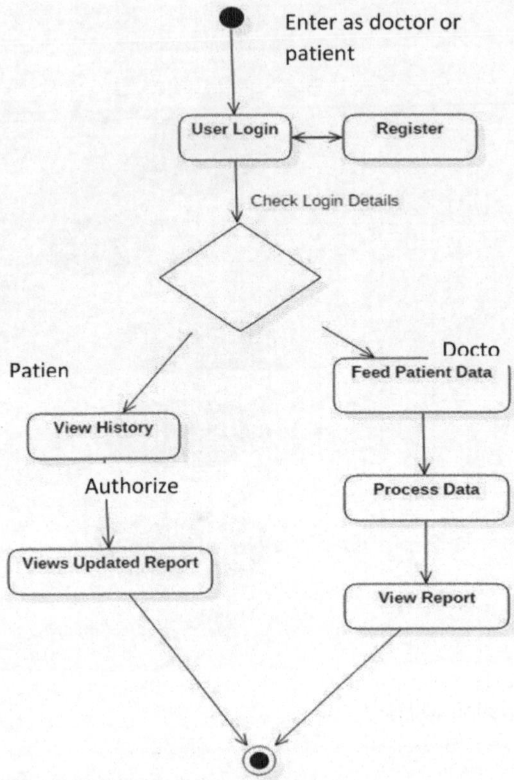


**FIGURE 10.9B** Various decision paths during system activity.

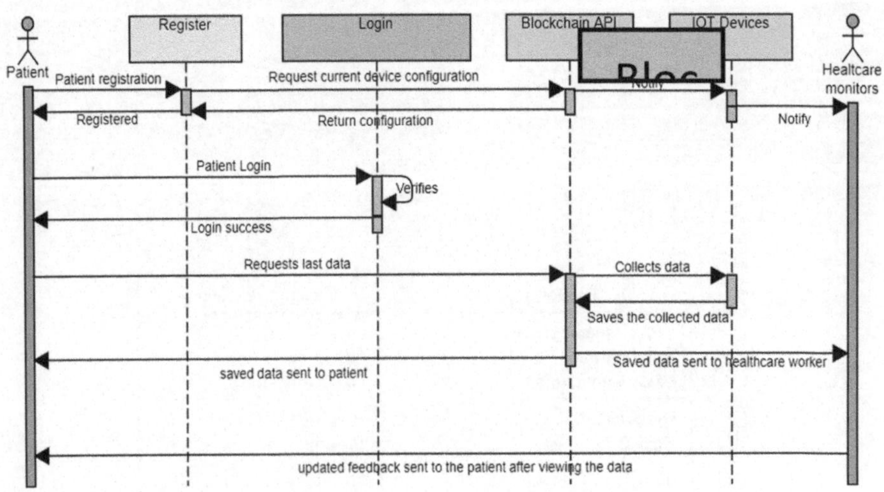


**FIGURE 10.10** Sequence model.

updated by thc healthcare worker, and that each time a record is added, and it is stored in the blockchain.

# MODEL OPERATION

## Design of the Database

### ER Diagram

The entity relationship (ER) diagram of the researchers' model (Figure 10.11a) shows the relationship between the patient, healthcare worker, and blockchain entities. The diagram also includes the attributes for each entity. The patient and healthcare worker can store as well as request data from the blockchain.

### Relational Model

Figure 10.11b represents how each patient record is stored. The blockchain can store numerous patient records as well as the updated healthcare records of the patient.

### User Interface

Figure 10.12 shows the landing page for the users.

The patient can give permission to the doctor to add a record. The doctor can add records as shown in Figure 10.13 or create a new record as in Figure 10.14. The patient can view the record as in Figure 10.15.

## Middleware

- Ethereum: It is an open-source blockchain that has smart contract functionality and is also decentralized. It helps in enabling distributed applications and smart contracts which could be used to be built and run without any interference from a third party, downtime, or fraud.
- Ganache: It is a tool that is used for creating your own blockchain for distributed app development. Ganache can be used during the whole life cycle, which helps in developing, deploying, and testing any decentralized apps in a safe environment.
- Truffle: It is mainly used for contract development, for testing blockchain environments, and is also an asset pipeline that is used for Ethereum blockchain. It provides a development framework for creating, designing, and deploying smart contracts for Ethereum.
- Visual Studio Code: It is a freeware source-code editor made by Microsoft for Linux, Windows, and macOS. It has numerous features, including syntax highlighting, debugging, code refactoring, embedded Git, intelligent code completion, and snippets.
- Node.js: It is an open environment used for the development of web applications. Node.js applications have a library that helps in developing web-based applications.

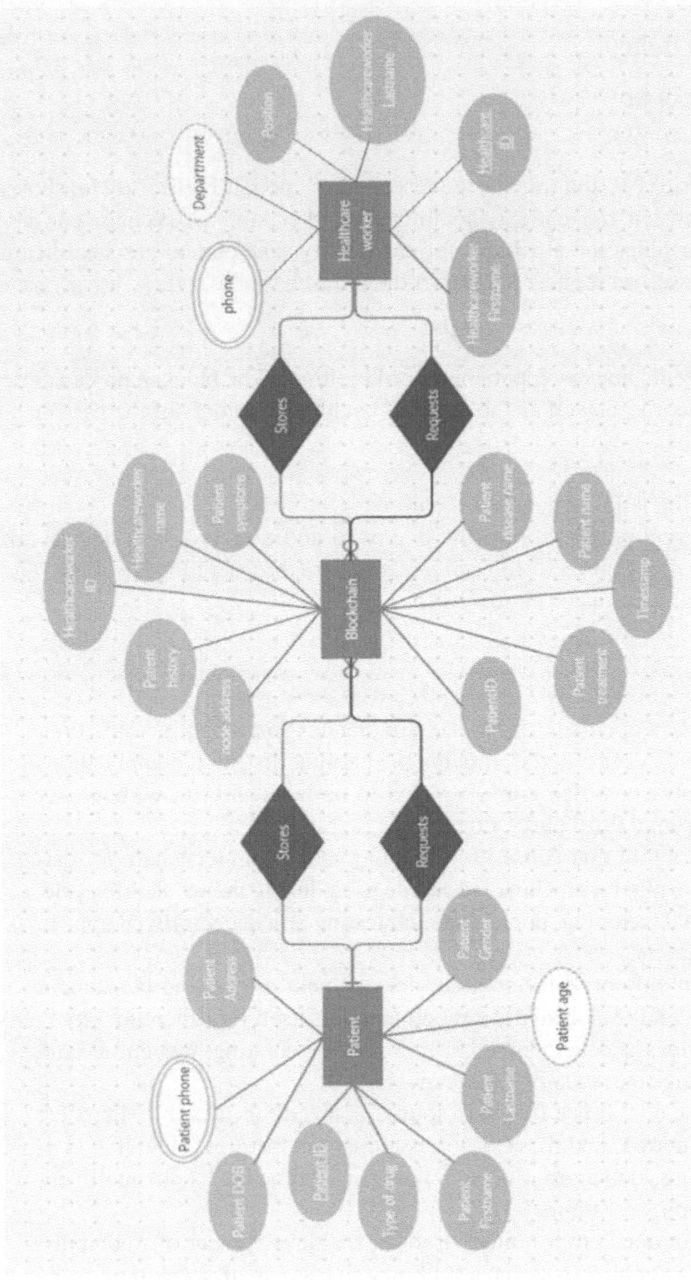


**FIGURE 10.11A** ER diagram.

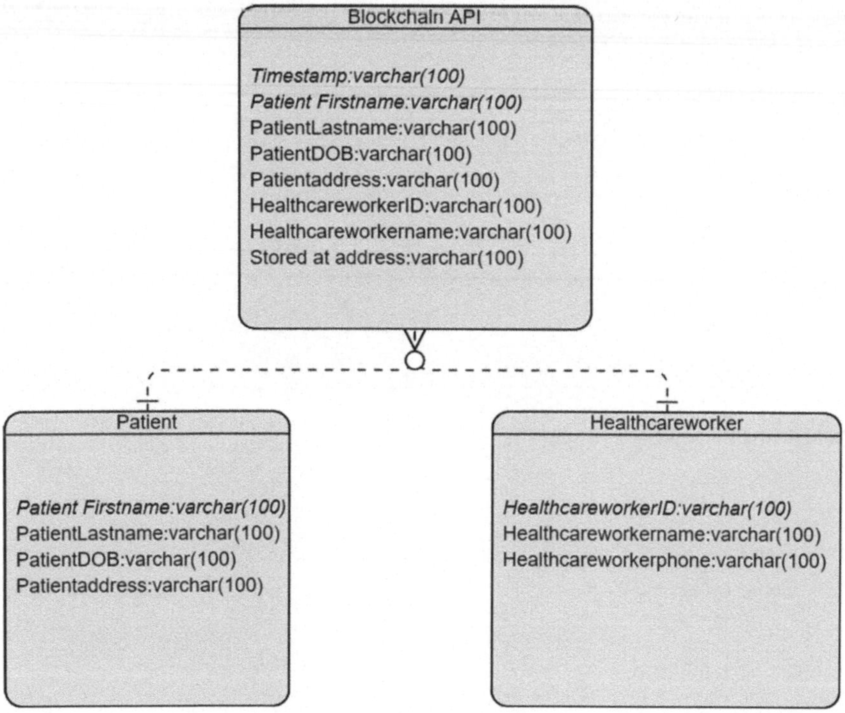


**FIGURE 10.11B** Relational model.

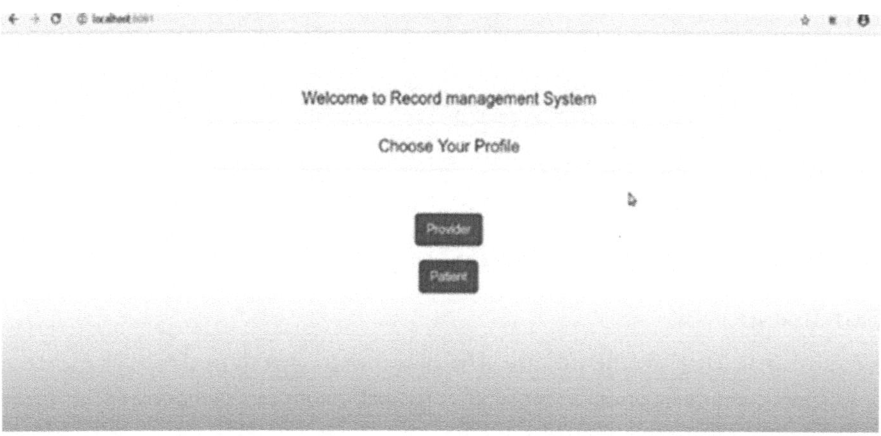


**FIGURE 10.12** User interface.

- Web3: It has a lot of libraries that allow you to work with nodes in Ethereum. The nodes can be either local Ethereum nodes or remote Ethereum nodes of the contract.

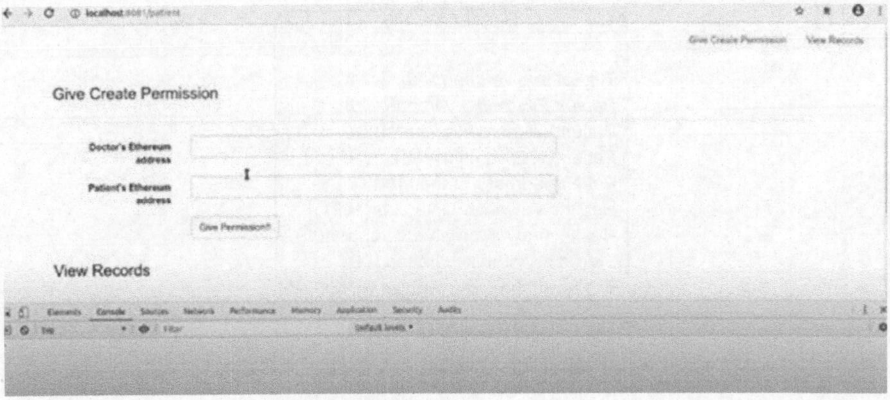


**FIGURE 10.13** Adding a patient's record.

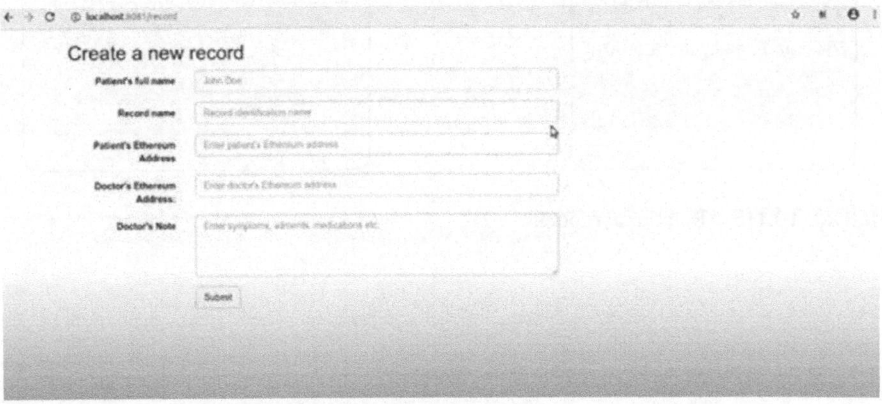


**FIGURE 10.14** Create new record.

# VERIFICATION AND VALIDATION RESULTS

## Unit Testing

Patients can be successfully registered as shown in Figure 10.16a.

Doctors can add a patient's record after the patient gives permission to the doctor as in Figure 10.16b.

Patients can view the added record as in Figure 10.16c.

**FIGURE 10.15** View records.

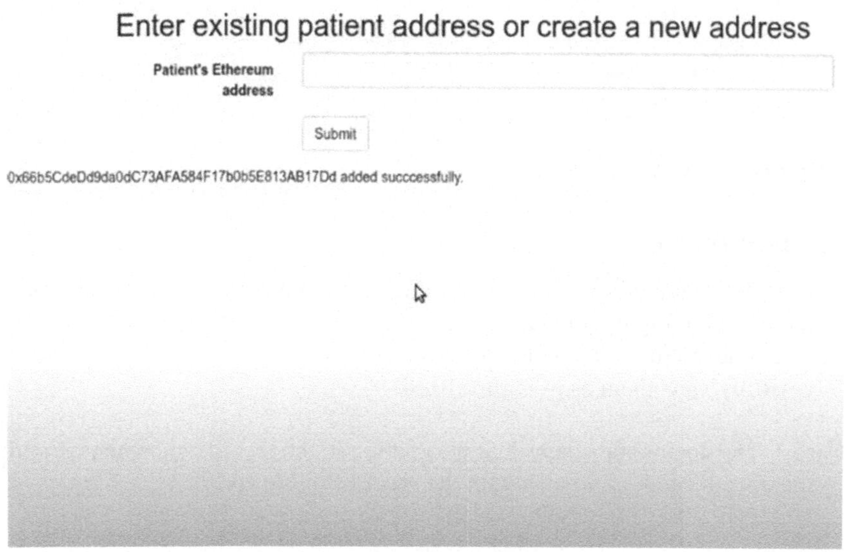


**FIGURE 10.16A** Register patients.

**FIGURE 10.16B** Create record.

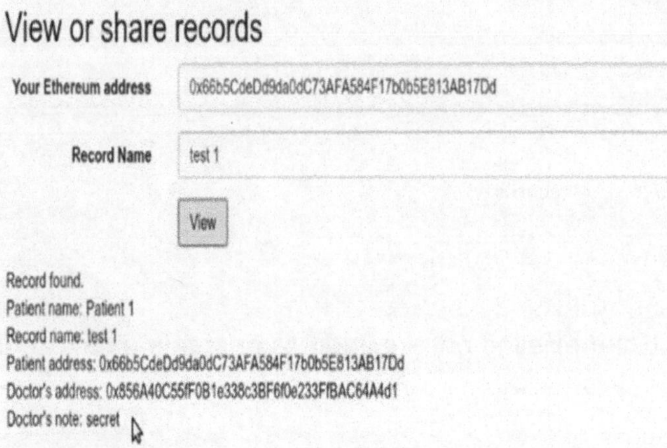


**FIGURE 10.16C** View records.

## Integration Testing

Case 1: The patient is registered and doctor can request to add the patient's records as in Figure 10.17a.

Case 2: The doctor adds the records, and the patient (Figure 10.17b) can successfully view them as in Figure 10.18.

Case 3: The doctor adds their digital signature to the record, and when viewed the doctor's address is seen with the record (Figure 10.19).

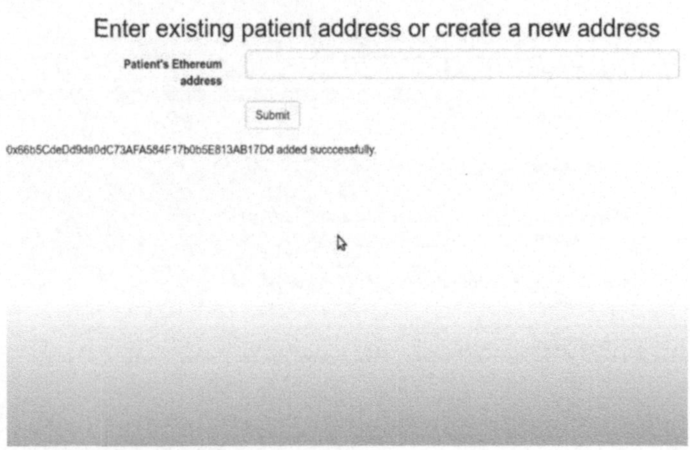


**FIGURE 10.17A** Register patients.

**FIGURE 10.17B** Register patients.

**FIGURE 10.18** View records.

## User Testing

The home page where users can choose the provider/patient is shown in Figure 10.20.

The doctor can create a new record for a patient with permission as in Figure 10.21.

Patients can give permission and view records as in Figure 10.22.

Patients can share their records as in Figure 10.23.

## Defect Analysis

A fishbone diagram (Figure 10.24) depicts the cause–effect analysis of defects.

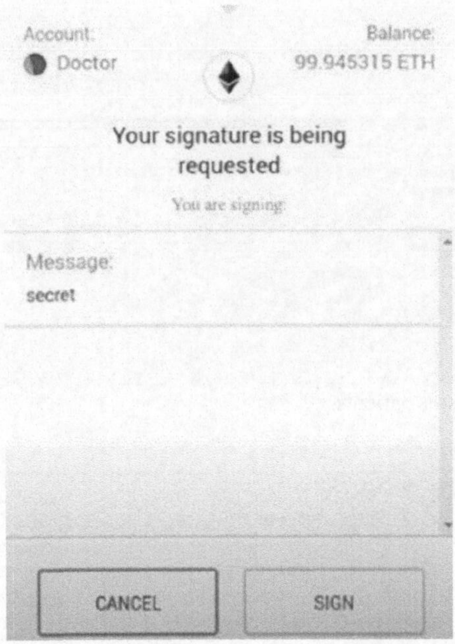


**FIGURE 10.19** Add digital signature.

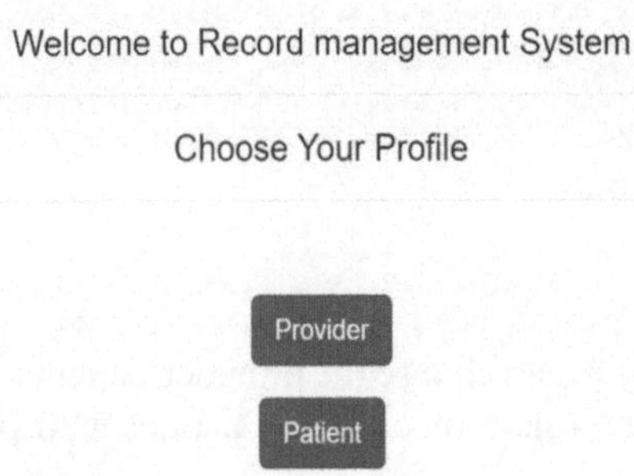


**FIGURE 10.20** Home page.

## McCall's Quality Factors

### Product Operation Factors

*Correctness*

The output of the modules of the project is accurate. However, the completeness of information shown depends on the input data.

*Reliability*

The system can have a failure rate of 0.1 per hour and so a higher reliability. In case of a service failure, the data and accounts are not lost and can be retrieved.

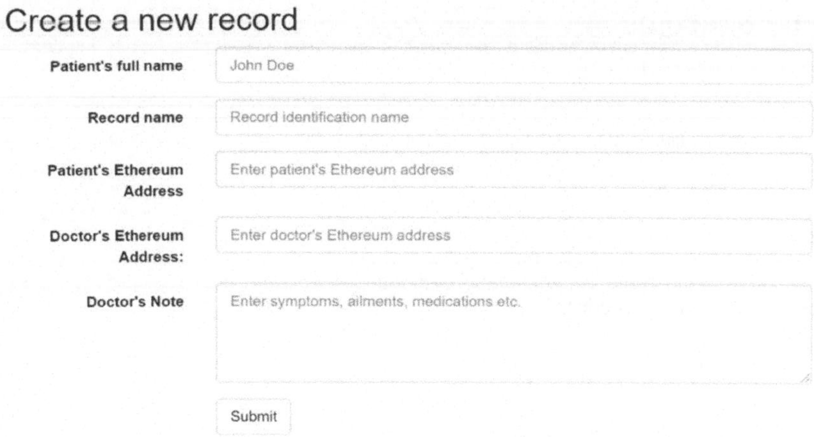


**FIGURE 10.21** Create new record.

**FIGURE 10.22** Give permission and view records.

*Efficiency*

For efficient working of the project, a system with a minimum processing speed of 2.2 GHz and 4GB RAM is required.

*Integrity*

The integrity of the system security is maintained by the encryption of data done by blockchain and the digital signatures that are added to each transaction to verify which user has committed the transaction.

FIGURE 10.23 Share records.

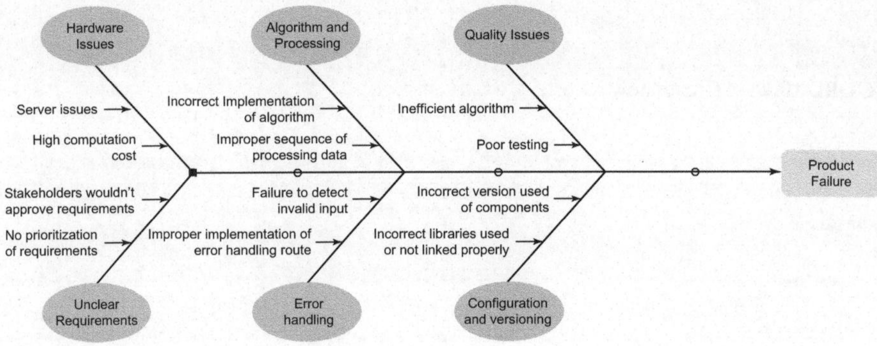


FIGURE 10.24 Fishbone diagram.

FIGURE 10.25 Add digital signature.

*Usability*

The project requires someone with a basic understanding of blockchain technology to use it, and higher knowledge for someone to make changes to it. Its usability is moderately easy and requires some prior knowledge of blockchain. The system also displays the interoperability of blockchain technology with web applications and databases.

## EXPERIMENT RESULTS AND ANALYSIS

### Results

The doctor adds their digital signature to the record (Figure 10.25).

The address of the doctor appears with the record when the patient views them as in Figure 10.26.

No one else can view the record without permission as in Figure 10.27.

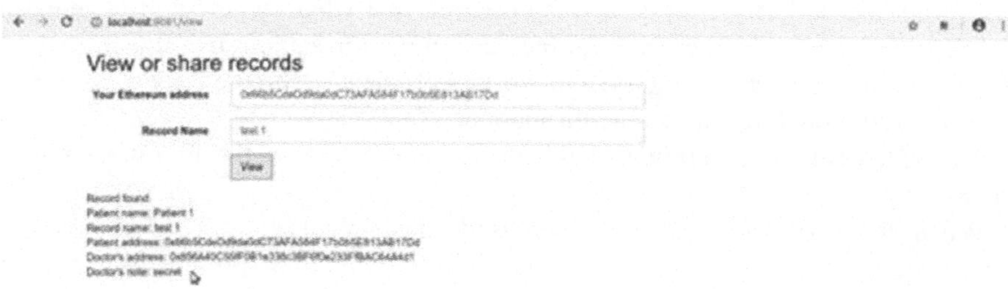


**FIGURE 10.26** Share records.

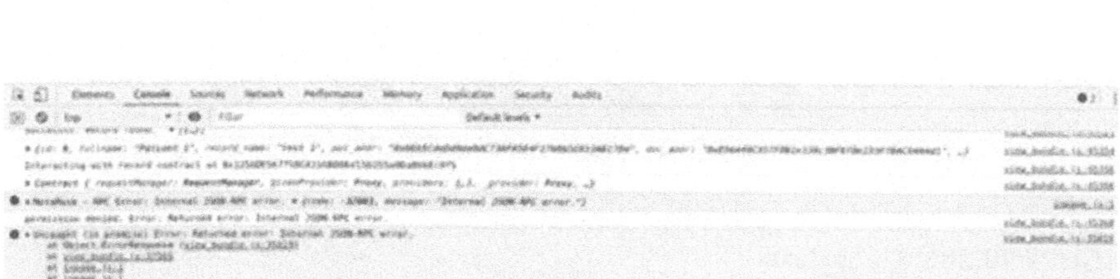


**FIGURE 10.27** No viewing without permission.

## RESULT ANALYSIS

The primary aim of the research is to provide a secure system that is the most trustworthy and if the data stored in the model is dependable and cannot be changed by anyone else. This blockchain-based system makes sure that the model maintains this method. Data that is stored in the model can only be changed by an authorized person. Other than that, the record data can be accessed only by doctors and patients. Doctors, patients, or any other person in the system cannot make modifications in the smart contract unless they have access to it. This makes sure that the information is private or the record of the patient cannot be accessed by everyone and stays tamper-proof. The private details of the patient are stored and a hash is created for the stored patient information. This hash is now stored and can be used when healthcare workers and patients need it. The hash is secured cryptographically. This is used to maintain security in our proposed system.

Performance-wise, we used Apache JMeter to evaluate our system. By using JMeter we simulated 100–500 users (with a period of 10 to 35 minutes) of the model using the system's different functions. The throughput of the JMeter is defined as Data/time, that is KB/sec units. It was observed that when the amount of requests and the users increased, there was an increase in the throughput in a linear manner. Hence, results showed that the system worked efficiently. Simulations were run on our proposed model to analyze the throughput.

The graphical representation of the results are shown in the following Figure 10.28.

When the transaction number increased, the execution time of the system increased. These transactions are for the different functions present in the smart contracts. When a single user uses the system, the functions for registering patient details, adding patient medical records, and viewing patient medical records would take 2.29, 4.3, and 3 seconds, respectively. The execution times increase when 100 users use the system at the same time.

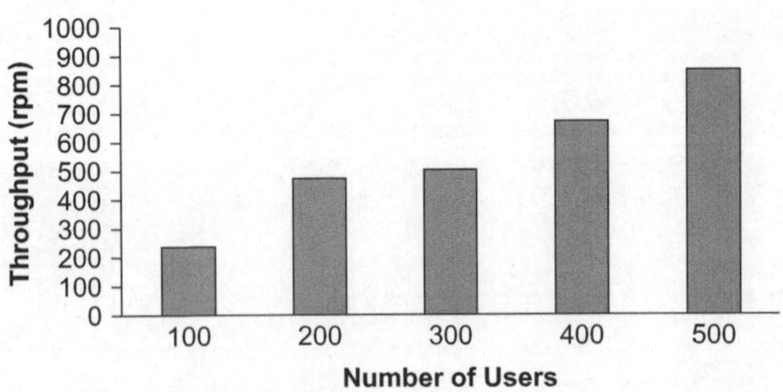


**FIGURE 10.28** Throughput.

## CONCLUSION AND FUTURE WORK

This blockchain framework can be used by healthcare facilities to overcome the limitations of the current way of storing data. Blockchain technology is the key to enhancing privacy and helps in preventing data breaches. The sharing of patient data between the concerned parties can be done in a safe and secure manner. Blockchain is the stepping stone to a more transparent healthcare industry. Thus, blockchain helps revolutionize healthcare sectors by providing privacy, accessibility and trust to medical records. The use of blockchain can further help in implementing a payment module in our existing blockchain structure. To implement, the researchers have to decide on the amount a patient would be willing to pay to consult a doctor in a decentralized model in blockchain. Certain policies and rules related to healthcare divisions also need to be gathered.

## REFERENCES

1. Ismail Keshta, Ammar Odeh, "Security and privacy of electronic health records: Concerns and challenges", *Egypt. Inform. J.*, vol. 22, no. 2, pp. 177–183, 2021. ISSN 1110-8665
2. Noshina Tariq, Ayesha Qamar, Muhammad Asim, Farrukh Aslam Khan, "Blockchain and Smart Healthcare Security: A Survey", *Procedia Com. Sci.*, vol. 175, pp. 615–620, 2020.
3. C. Agbo, Q Mahmoud, J Eklund, "Blockchain technology in healthcare: A systematic review", *Healthcare (Switzerland)*, vol. 7, no. 2, p. 56, 2019.
4. M. Attaran, "Blockchain technology in healthcare: Challenges and opportunities", *Int. J. Health. Manag.*, vol. 15, no. 1, pp. 70–83, 2022.
5. Tomaso Aste, Paolo Tasca, Tiziana Di Matteo, "Blockchain technologies: The foreseeable impact on society and industry", *Computer*, vol. 50, no. 9, pp. 18–28, 2017.
6. Naipeng Dong, Hugo Jonker, Jun Pang, "Challenges in eHealth: From enabling to enforcing privacy", 2018. https://doi.org/10.1007/978-3-642-32355-3_12
7. W. J. Gordon, C. Catalini, "Blockchain technology for healthcare: Facilitating the transition to patient-driven interoperability", *Comput. Struct. Biotechnol. J.*, vol. 16, pp. 224–230, 2018.
8. L. Coventry, D. Branley, "Cybersecurity in healthcare: A narrative review of trends, threats and ways forward," *Maturitas*, vol. 113, pp. 48–52, 2018.
9. M. Reisman, "EHRs: The challenge of making electronic data usable and interoperable," *P T*, vol. 42, no. 9, pp. 572–575, 2017.
10. S. T. Argaw, N. E. Bempong, B. Eshaya-Chauvin, A. Flahault, "The state of research on cyberattacks against hospitals and available best practice recommendations: A scoping review," *BMC Med. Inform. Decis. Mak.*, vol. 19, no. 1, pp. 1–11, 2019.
11. S. Nakamoto, "Bitcoin: A peer-to-peer electronic cash system," *SSRN Elect. J.*, pp. 1–9, 2008. https://doi.org/10.2139/ssrn.3977007
12. G. Jetley, H. Zhang, "Electronic health records in IS research: Quality issues, essential thresholds and remedial actions," *Decis. Support Syst.*, vol. 126, pp. 113–137, 2019.
13. A. Boonstra, A. Versluis, J. F. J. Vos, "Implementing electronic health records in hospitals: A systematic literature review," *BMC Health Serv. Res.*, vol. 14, no. 1, p. 370, 2014.
14. Z. Zheng, S. Xie, H. Dai, X. Chen, H. Wang, "An overview of blockchain technology: Architecture, consensus, and future trends", *Proc. - 2017 IEEE 6th Int. Congr. Big Data, BigData Congr*, Vol. 2017, pp. 557–564, 2017.

15. C. Pirtle, J. Ehrenfeld, “Blockchain for healthcare: The next generation of medical records?”, *J. Med. Syst.*, vol. 42, no. 9, p. 172, 2018.
16. A. A. Siyal, A. Z. Junejo, M. Zawish, K. Ahmed, A. Khalil, G. Soursou, “Applications of blockchain technology in medicine and healthcare: Challenges and future perspectives,” *Cryptography*, vol. 3, no. 1, p. 3, 2019.
17. D. Vujičić, D. Jagodić, S. Randić, “Blockchain technology, bitcoin, and Ethereum: A brief overview,” *2018 17th Int. Symp. Infoteh-Jahorina, Infoteh 2018 - Proc.*, vol. 2018, pp. 1–6, 2018.
18. S. Wang, Y. Yuan, X. Wang, J. Li, R. Qin, F. Y. Wang, “An overview of smart contract: Architecture, applications, and future trends,” *IEEE Intell. Veh. Symp. Proc.*, vol. 2018, no. iv, pp. 108–113, 2018.
19. T. T. Kuo, H. E. Kim, L. Ohno-Machado, “Blockchain distributed ledger technologies for biomedical and health care applications”, *J. Am. Med. Informatics Assoc.*, vol. 24, no. 6, pp. 1211–1220, 2017.
20. P. Zhang, J. White, D. C. Schmidt, G. Lenz, S. T. Rosenbloom, “FHIRChain: Applying blockchain to securely and scalably share clinical data,” *Comput. Struct. Biotechnol. J.*, vol. 16, pp. 267–278, 2018.
21. S. Gupta, M. Sadoghi, “Blockchain transaction processing,” *Encycl. Big Data Technol.*, pp. 366–376, 2019.
22. G. Wood, “Ethereum: A secure decentralised generalised transaction ledger. EIP-150 REVISION,” 2017, no. August 1, 2017, p. 33, 2017.
23. N. Atzei, M. Bartoletti, T. Cimoli, S. Lande, R. Zunino, “SoK: Unraveling bitcoin smart contracts,” in *7th International Conference, POST 2018 Held as Part of European Joint Conferences on Theory and Practice of Software, ETAPS 2018 Thessaloniki, Greece*, 2018, pp. 217–242.
24. I. Grishchenko, M. Maffei, C. Schneidewind, “A semantic framework for the security analysis of ethereum smart contracts,” in *Principles of Security and Trust*, 2018, pp. 243–269.
25. Niranjanamurthy, K. Kumar S, A. Saha, D. D. Chahar, “Comparative study on performance testing with JMeter,” *Int. J. Adv. Res. Comput. Commun. Eng.*, vol. 5, no. 2, pp. 70–76, 2016.

# Index

## C



## D



## E

## K



## L



## M



## N



## O

## P



## Q



## R



## S